Smart Grid (R)Evolution

The term "smart grid" has become a catch-all phrase to represent the potential benefits of a revamped and more sophisticated electricity system that can fulfill several societal expectations related to enhanced energy efficiency and sustainability. Smart grid promises to enable improved energy management by utilities and by consumers, to provide the ability to integrate higher levels of variable renewable energy into the electric grid, to support the development of microgrids, and to engage citizens in energy management. However, it also comes with potential pitfalls, such as increased cybersecurity vulnerabilities and privacy risks. Although discussions about smart grid have been dominated by consideration of technical and economic dimensions, this book takes a sociotechnical systems perspective to explore critical questions shaping energy system transitions. It will be invaluable for advanced students, academic researchers, and energy professionals in a wide range of disciplines, including energy studies, environmental and energy policy, environmental science, sustainability science, and electrical and environmental engineering.

JENNIE C. STEPHENS is an Associate Professor and the Blittersdorf Professor of Sustainability Science and Policy at the University of Vermont's Rubenstein School of Environment and Natural Resources and its College of Engineering and Mathematical Sciences. Jennie's research, teaching, and community engagement focus on the sociopolitical aspects of energy technology innovation, electricity system change, and climate change communication. She has contributed to the understanding of the social dynamics of wind power, carbon capture and storage, and smart grid, and she brings to this project experience in stakeholder engagement and communication with experts, academics, and the public. She is particularly interested in facilitating social learning as we transition away from fossil fuel–based energy systems toward renewables-based systems.

ELIZABETH J. WILSON is Associate Professor of Energy and Environmental Policy and Law at the Humphrey School of Public Affairs and a Fellow at the Institute on the Environment, both at the University of Minnesota. Elizabeth brings to this project extensive knowledge on the importance of subnational factors in shaping energy policy and technology deployment, with a special focus on policies, regulatory and legal frameworks for emerging energy technologies in states, and Regional Transmission Organizations. Her research has also focused on regulatory and governance systems for carbon capture and sequestration, energy efficiency, and wind power.

TARLA RAI PETERSON is Professor of Environmental Communication at the Swedish University of Agricultural Sciences and holds the Boone and Crockett Chair of Wildlife and Conservation Policy at Texas A&M University. Tarla Rai has published several books and numerous articles on the intersections between communication, policy, and democratic practice, particularly as these intersections relate to science and technology. Her most recent book, *The Housing Bomb*, explores these intersections as they relate to public participation in development of policy that could enhance sustainability. She brings to this project extensive theoretical and critical expertise on public perceptions of environmental issues and technologies and the critical role of both mediated and interpersonal communication in the practice of democracy among members of a pluralistic society.

SMART GRID (R)EVOLUTION

Electric Power Struggles

Jennie C. Stephens
University of Vermont

Elizabeth J. Wilson
University of Minnesota

Tarla Rai Peterson
Swedish University of Agricultural Sciences

CAMBRIDGE
UNIVERSITY PRESS

CAMBRIDGE
UNIVERSITY PRESS

32 Avenue of the Americas, New York NY 10013-2473, USA

Cambridge University Press is part of the University of Cambridge.

It furthers the University's mission by disseminating knowledge in the pursuit of
education, learning and research at the highest international levels of excellence.

www.cambridge.org
Information on this title: www.cambridge.org/9781107635296

© Jennie C. Stephens, Elizabeth J. Wilson, and Tarla Rai Peterson 2015

First published 2015
First paperback edition 2016

A catalogue record for this publication is available from the British Library

Library of Congress Cataloguing in Publication data
Stephens, Jennie.
Smart grid (r)evolution : electric power struggles / Jennie C. Stephens,
University of Vermont, Elizabeth J. Wilson, University of Minnesota,
Tarla Rai Peterson, Texas A&M University.
pages   cm
Includes bibliographical references.
ISBN 978-1-107-04728-0 (Hardback)
1. Smart power grids.   I. Wilson, Elizabeth J.   II. Peterson, Tarla Rai.
III. Title.   IV. Title: Smart grid evolution.   V. Title: Smart grid revolution.
TK3105.S74 2015
333.793´2–dc23   2014035119

ISBN 978-1-107-04728-0 Hardback
ISBN 978-1-107-63529-6 Paperback

We dedicate this book to our families

Contents

Foreword

What makes a book worth reading? Has it a *theme*, a topic of importance that matters? Perhaps we can use it as a *tool*, to do good things? Can it take us on a *journey* to new worlds? Will it *stretch our minds* and challenge the old thoughts in them? This book, **Smart Grid (R)evolution**, offers value by each of those measures.

If you want a grand and vital *theme*, here it is. No crisis is more all-threatening than climate-change, and nothing we can do about it is more important than controlling the generation of electricity, and no campaign to optimize generation of electricity is viable without:

smarter decisions about which power-plants to turn on and off,
smarter decisions about which transmission lines to open and close, and
smarter ways to show customers how their acts affect operations and costs:
in short, a smarter grid.

This book looks from several angles at the promises and pitfalls that lie between the grid systems of 'today' and the emergence of that smarter grid.

Is this book a *tool*? Yes; it's a multi-task, Swiss-Army-knife, kind of tool, with tips for readers on everything from listening to customers, to looking at data across-the-board, to balancing the pace of infrastructure investments among retail, wholesale, and operational installations and practices. Anyone charged with operating an electrical system, anyone concerned about using an electrical system, and anyone worried about paying for one, will find useful insights in this text.

Is it a *journey*? Yes; this story takes us from the sea-flooded subways of New York, to the mountains of Boulder, to Austin in the Texas plain, and to the North Sea cliffs of Bornholm. Around the world, the book shows us the early seeds and the emerging shoots of a radically new system. It is a journey over time as well as over miles, and from mind to mind, from group to group. It treats, with respect, the hopes of many and the concerns of others. So we are led to both promises and fears, with a calm and reasoned summary of each.

In Emily Dickinson's words, the authors 'tell the truth, but tell it slant," looking at reality and dreams from multiple angles. In their own analogy, they see disparate groups, each like the blind men touching different portions of an elephant, and they try to move us, together, past those unconnected and disparate views. Thus, they tell the emerging story of smart grid systems from multiple perspectives, with real respect for differing views, but without abandoning the authors' own judgments.

Do the stories and the analyses *stretch our minds* and challenge our old thoughts? Speaking for myself, I've spent three decades nurturing technology change, worrying about climate change, overseeing electric system operations and searching for positive value from disputes as I judged and resolved contested cases about new policies. With that basis – or despite it! – I saw new things here, shifted my weighing of some risk assessments, opened my eyes to the emotions of people with whom I might disagree on policy judgments and widened my sense of the possible in our future. I have to think that any serious reader will see things here that make them reach beyond their old beliefs.

How do the authors do this? Well, part of the answer is simple old-fashioned hard work, since **Smart Grid (R)evolution** is the result of years of gathering of information by three talented and complementary scholars, with a track record of producing good work together. But, its not *just* hard work that makes this book good. There is also an intellectual framework, a conceptual structure, of real merit. The authors' approach is not just technical or economic, but they also consider and present the social and political elements of reaching a social consensus. This process can be labeled 'socio-technical systems analysis." The label sounds arcane, but it reflects the deep roots and fundamental value of the approach used here.

The authors (like most of us) have emerged from an intellectual tradition that (since the Enlightenment) has increasingly treated knowledge as divided into what Lord Snow called 'Two Cultures, " one focused on literature and social understanding, and the other dedicated to technical and scientific rigor. Fortunately, the authors of this book recognize that few important questions are solely technical and few are solely social. A smart meter's real meaning emerges only as part of a smart grid, just as a smart phone's merit fades if not connected to a smart network. And a smart grid, like a smart network, requires co-ordination among human beings as much as it requires frequency regulation. More generally, the history of technical change illustrates the significance of what I have called 'the heaviness of existing reality." Moving past the heaviness of current investment requires thinking seriously, and sympathetically, about social and political issues as well as about engineering ones. Yet, the converse is also true; mere social consensus (or even political unanimity) about goals will not hold back an incoming tide or alter "an inconvenient truth." The only likely path forward is to blend social and technical analyses in ways that bridge the divide between "The Two Cultures," This book's deepest and most important strength is that it will help its readers do just that.

Michael Dworkin is Professor of Law and Director of the Institute for Energy and the Environment at Vermont Law School. He serves on the Boards of the Vermont Electric Power Company (VELCO) and of the Vermont Energy Investment Corporation (VEIC) and has been adjunct faculty for the University of Houston Law Center, the University of Waikato Center for Energy Resources and the Environment, and the Engineering & Public Policy Department of Carnegie Mellon University. In the past, he was Chairman of the Vermont Public Service Board, President of the New England Conference of Public Utilities and Member of the Executive Committee of the Electric Power Research Institute.

Acknowledgments

This book is the culmination of years of collaborative research that involved engagement, support, and contributions from many people.

First of all, we would like to acknowledge and thank the graduate research assistants without whom this research would not have happened, including Paulami Banerjee, Xiao Chen, Ryan Collins, Miriam Fischlein, Julia Eagles, Peter Frongillo, Cristi Horton, Drema Khraibani, Clark Koenigs, Ria Langheim, Joel Larson, Melissa Skubel, Adrienne Strubb, Mudita Suri, Sophia Ran Wang, Michelle Wenderlich, and Lauren Ziemer. Several undergraduate research assistants also contributed, including Kaitlin Dawson, Caroline Ellings, Emily Krieter, David Love, William Maxwell, Danielle Miller, Ramsey Randolph, and Noopur Shah.

Our families and partners have provided multiple different kinds of support throughout this process. We are so very appreciative of Dan Bolon, Cecelia Stephens Bolon, Anna Stephens Bolon, Ross Jackson, Amelia Wilson-Jackson, Charlie Wilson-Jackson, Judy Scott, Jim Scott, Bill Wilson, Jan Wilson, Markus Peterson, Nils Peterson, Wayne Peterson, and Scott Peterson. We also deeply thank Cathal and Sarah Stephens for welcoming us to stay in their net-zero energy home in Donegal, Ireland, where we began writing this book.

We are most grateful for the financial support for this research that we received from the National Science Foundation's Science, Technology and Society program (NSF-SES 1127697, NSF-SES 1316442, NSF-SES-1127272, NSF-SES 1316330, NSF-SES 112760, NSF-SES 1316605). We are particularly appreciative of the support and encouragement we received from our NSF program officers Linda Layne and Kelly Moore. We have also received institutional support from Clark University's Marsh Institute on Human Dimensions of Global-Environmental Change, Clark's Mosakowski Institute, UVM's Rubenstein School of Environment and Natural Resources and the College of Engineering and Mathematical Sciences, the Humphrey School of Public Policy and the Institute of the Environment at the University of Minnesota, and the Environmental Communication Program at the Swedish University of Agricultural Sciences. We appreciate all of the support we have received to expand energy research and energy education beyond the technical and engineering details.

For invaluable guidance and perspective throughout the duration of this project, we thank our project advisory board, Massoud Amin, Jay Apt, and Halina Brown, who gave generously of their time and ideas.

Other academic colleagues that have also contributed in different direct and indirect ways include Chuck Agosta, Jonn Axsen, Nick Belanger, Sally Benson, Hanna Bergeå, Seth Blumsack, Mary-Ellen Boyle, Mike Bull, Nancy Budwig, Ann Carlson, Anabela Carvalho, Xavier Deschenes-Philion, Sairaj Dhople, Pamela Dunkle, Danielle Endres, Sabine Erlinghagen, Jim Gomes, Rob Goble, Andrea Feldpausch-Parker, Cristián Alarcón Ferrari, Hans Peter Hansen, Maya Jegen, Scott Jiusto, Alexandra Klass, Mladen Kezunovic, Jennifer Kuzma, Alexandra Mallett, Jochen Markhard, Steve McCauley, James Meadowcroft, Granger Morgan, Natalie Nelson-Marsh, Hari Osofsky, Derek Peters, Melisa Pollak, Ryan Reiber, Ian Rowlands, Jodi Sandfort, Rebecca Slayton, Pete Seiler, Tim Smith, David Solan, Nadarajah Sriskandarajah, Margaret Taylor, Glen Toner, Shalini Vajjhala, Philip Vergragt, Mark Winfield, and Bruce Wollenberg. Our collaborative meetings with our Canadian colleagues have been particularly influential.

A transdisciplinary book such as this one requires its authors to draw from a broad spectrum of expertise that spans both academic disciplines and societal organizations. Numerous individuals and organizations were generous with their time. We thank the following organizations for their willingness to participate in focus groups and/or interviews: ERCOT, MISO, ISO-NE, CAISO, Mass Energy Consumers Alliance, Worcester Polytechnic Institute, Mass Energy, Clark University, Shrewsbury Electric and Cable Operations, Worcester Housing, Energy and Community Group (WOHEC), MA DPU, National Grid, Great Plains Institute, MN State Energy Office, Xcel Energy, Great River Energy, University of Minnesota, Oncor, College Station Utilities, Texas A&M, Austin Energy, Texas PUC, OPUC, Center Point Energy, Conservation Law Foundation, Sierra Club, Environmental Defense, Natural Resources Defense Council, Vermont Law School, Vermont Natural Resource Council, Vermont Electric Coop, Vermont Department of Public Service, University of Vermont Smart Grid IGERT Program, Green Mountain Power Energy Innovation Center, Vermont Public Interest Group, The Utility Reform Network (TURN), Long Island Power Authority, New York State Energy Research and Development Authority, New York Power Authority, New York Independent System Operator, New York Energy Consumers Council, New York State Public Service Commission, Bonneville Power Association, California Public Utilities Commission, California Energy Commission, Sacramento Municipality Utility District, and Pacific Gas and Electric, and Østkraft (the distribution system operator on Bornholm, Denmark).

We would also like to thank several specific individuals who provided informal advice and assistance multiple times throughout the project, including Mike

Gregerson from the Great Plains Institute, Matt Ellis and Rao Konidena from MISO, Anthony Giaconomi from ISO-New England, and Paul Wattles from ERCOT. We have had the opportunity to interact with many wonderful people throughout the years and feel fortunate to have been able to meet, engage, and work with kind and insightful professionals. While they have been generous with their time and knowledge, all errors that remain in this manuscript are the authors' alone.

Tables

Figures

Acronyms

AC	Alternating Current
AMI	Advanced Meter Infrastructure
AMR	Automatic Meter Reader
ARRA	American Recovery and Reinvestment Act
CAES	Compressed Air Storage
CO_2	Carbon Dioxide
C-BED	Community-Based Energy Development
CREZ	Competitive Renewable Energy Zones
DC	Direct Current
DG	Distributed Generation
DER	Distrbuted Energy Resources
DIR	Dispatchable Intermittent Resources
DoD	Department of Defense
DoE	Department of Energy
DSM	Demand Side Management
EDF	Environmental Defense Fund
EMF	Electromagnetic Fields
EMS	Energy Management System
EPA	Environmental Protection Agency
ERCOT	Electricity Reliability Council of Texas
EU	European Union
ESCO	Energy Service Companies
FACTS	Flexible AC Transmission System
FERC	Federal Energy Regulatory Commission
GIS	Geographic Information Science
GTI	Grid-tie Inverter
HVAC	Heating, Ventilation, and Air Conditioning
kV	kilo Volts
kWh	kilowatt hour

ICT	Information Communication Technology
ISO	Independent System Operator
MISO	Midcontinent Independent System Operator
MW	Megawatt
NASA	National Aeronautics and Space Administration
NE-ISO	New England Independent System Operator
NRDC	Natural Resources Defense Council
NREL	National Renewable Energy Lab
OLTS	On-Load Tap-Charger
OMS	Outage Management System
PEV	Plug-in Electric Vehicle
PG&E	Pacific Gas and Electric
PHEV	Plug-in Hybrid Electric Vehicle
PTC	Production Tax Credit
PUC	Public Utilities Commission
PUCT	Public Utility Commission of Texas
PURPA	Public Utilities Regulatory Policy Act
PV	Photovoltaic
R&D	Research and Development
REC	Renewable Energy Credit
RF	Radio Frequency
RPS	Renewable Portfolio Standard
RTO	Regional Transmission Organization
SCADA	Supervisory Control and Data Acquisition
SDG&E	San Diego Gas & Electric
SMUD	Sacramento Municipal Utility District
SPIDERS	Smart Power Infrastructure Demonstration for Energy Reliability and Security
SVC	Static Var Compensation
UMTDI	Upper Midwest Transmission Development Initiative
var	Volt-Ampere Reactive

1

Emerging Smart Grid Struggles

1.1 Vulnerability and Change

At the end of October 2012, more than eight million homes lost power as Superstorm Sandy battered the east coast of the United States. The electricity system disruption from this extreme weather event impacted households and businesses across seventeen states, including those as far west as Michigan. The storm left some without power for weeks, and lower Manhattan was in the dark for several days. The storm closed the New York Stock Exchange, the most powerful market in the world, for two full days, and Broadway shows were canceled for three consecutive days (Webley 2012). The storm forced evacuation of critical care patients and premature babies at New York University's hospital to another hospital in the dark, and flooded substations and downed power lines caused unprecedented levels of disruption to the city's energy systems. As the disaster unfolded, the vulnerabilities of the United States' electric system were broadcast to the world.

In the aftermath of the storm, the region struggled to recover and restore electricity. In some places, the same vulnerable electricity system, with the same basic technologies and same structure, was reinstalled, demonstrating a common and fundamental irony of disaster recovery. Although a disruption provides a window of opportunity to upgrade technology and introduce new approaches to enhance system resilience, established policies and procedures often require investment in and installation of the same infrastructure. But Hurricane Sandy also sparked broad societal discussion on the vulnerability of the electric infrastructure and has encouraged long-term plans and investments to improve reliability and resilience. Investment decisions after a major disruption represent one of many emerging struggles of electricity system change. When the Secretary of the United States Department of Energy, Ernie Moniz, spoke about Hurricane Sandy in his first major policy speech in August 2013, he said: "we have to help this rebuilding in a smart way" (Moniz 2013). In this political statement, Moniz was underscoring this critical challenge in electricity system change. When system disruptions occur, the pressure to "get the lights back on" surpasses all else. A clear tension exists between the immediate need to recover from an outage and the longer-term need for changes to move toward a future system that is more reliable and resilient. Rarely have electric utilities been able to use outage and system disruptions as opportunities to upgrade and update their technologies.

The aftermath of Superstorm Sandy highlights other struggles associated with change in electricity systems. Not only are there limits to the introduction of new technologies during a disruptive event, but also, overlapping jurisdictions and diverse priorities and perspectives among actors make electricity system change extremely challenging. Change in all complex social and technical systems is dependent on struggle and tension, and conflict creates possibilities for new and creative socio-technical norms to emerge. As we confront the challenges and opportunities of electricity system change, understanding how struggles are developing and why tensions are evolving can contribute to creative alignment of interests and priorities.

Multiple tensions and opportunities are currently emerging in electricity systems as the notion of a smarter grid offers both great promise and pitfalls. The term *smart grid* has become a catch-all phrase to represent the potential benefits of a revamped and more sophisticated electricity system that can fulfill several societal expectations related to enhanced efficiency and sustainability. Smart grid is not a single technology but a somewhat ambiguous term that represents multiple visions and technologies throughout the electric system. Smart grid often means different things to different people. Given the breadth of the many promises (and pitfalls) of smart grid, given the complexity of possible technical configurations of smart grid, and given the diversity of societal actors involved and invested in smart grid deployment, understanding the sociotechnical context for smart grid development is challenging and complicated.

Acknowledging the very different perspectives and priorities of the individuals and organizations involved in and impacted by electricity system change, this book explores and explains the dynamic smart grid landscape, exploring how new tensions create opportunities for evolutionary change and the potential for revolutionary change. In this book, we take a broad system-wide perspective to examine the different ways smart grid is meeting the evolving demands of electricity systems. By comparing smart grid development in different regions of the United States and Europe, we demonstrate that how smart grid is fulfilling changing societal expect-ations of electricity systems depends on social and political contexts, which are often shaped by regionally specific goals, resources, and engaged actors. Which actors and organizations have control and influence over shaping smart grid, and who benefits from smart grid changes, varies considerably among communities, states, regions, and countries.

Different smart grid visions reflect a diversity of social and political landscapes creating an evolving patchwork of smart grid trajectories. The diversity embedded within smart grid visions reflects a new diverse reality for energy systems, energy policy, and energy technologies. There is no silver-bullet solution to the energy challenges facing society. By revealing the diversity of smart grid potentials in this book, we also reveal and showcase the critical importance of context-specific approaches to considering energy system change.

The multifaceted diversity of perceptions of smart grid makes the Indian parable of the elephant and the blind men a useful analogy to understand our goals for this book. This story is often used to demonstrate how any individual's subjective experience may be accurate or true to themselves, but an individual's capacity to know the full truth remains limited. In the parable, six blind men touch different parts of an elephant, experiencing distinctly different realities. The man who touches the elephant's trunk feels a long, strong, thin animal, while the men who touch the hind leg, the tusks, the underbelly, and the tail each experience and envision a very different animal. Each of these individuals' perceptions is informed by the parts of the whole that they experience, but each man has little capacity to understand and interpret the full magnitude or shape of the whole elephant.

Such is the case with smart grid. Just as each blind man experiencing different parts of a large animal is unable to understand the entire elephant, different actors involved in and influenced by smart grid development are each engaged with different parts of the electricity system and have only a limited perspective on the potential and challenges of broad electricity system change. Each individual or organization is able to view only a limited part of the entire system. And the lack of a comprehensive educational approach to energy and electricity systems perpetuates this piecemeal understanding.

The insights we share in this book are built on six years of research that involved talking with hundreds of different people who are engaged in shaping electricity systems. As we listened and learned from people representing a wide range of organizations, from grid operators in the Midwest to small cooperative utilities in Vermont, we attempted to integrate different perspectives and priorities of smart grid with an ultimate goal of understanding the complexity of change and evolution in electricity systems. In this synthesis of our research, we attempt to step back far enough to enable readers to see the entire smart grid animal, but also provide sufficient detail for those who are especially interested in specific components.

1.2 The Grid Matters! Why We Care

Most people do not think much about electricity systems. People pay attention to "the grid" only when the power goes out or when a monthly bill is due. System reliability and affordability have been major tenets shaping electric system development for the past 150 years. While the first electric systems focused solely on powering lights, electricity systems have become increasingly critical infrastructure. More than ever before, we rely on electricity for communication, food, health, transportation, and other basic needs. If the power goes out we quickly become paralyzed when we are unable to charge our cell phones, pay our bills, refrigerate our food, and run our businesses and households.

Concern about the social impact of power outages and connections between electricity system vulnerabilities and more intense and frequent extreme weather events has been growing in the United States, as highlighted by Superstorm Sandy. Despite growing acknowledgement of the need to enhance resilience of the electric grid, investment in U.S. infrastructure is low, prompting the American Society of Civil Engineers to assess the U.S. energy system with a grade of D+ on their infrastructure report card (ASCE 2013).

Similar concerns are mounting in Europe, but the debates differ significantly among countries. Some European countries have a higher level of political and societal support and expectation for investing in infrastructure maintenance, so this alters the landscape for considering electricity system change. For example, in Germany, a national-level commitment to transitioning to a renewables-based electricity system has highlighted challenges regarding long-distance transmission planning. Microgrid planning in Denmark addresses system resilience as well as environmental concerns. In Italy, smart meter installations were initially driven by a desire to address electricity theft. In both the United States and Europe, preparing the grid to become more resilient to disruptions has become one of several motivations for growing interest in smart grid innovation.

The term smart grid does not have a precise, uniformly accepted definition. Rather, it is a vague, politically attractive, seemingly benign, and somewhat ambiguous phrase. After all, who would argue for a "dumb grid?" It is an umbrella term that encompasses many different technical and social changes affecting the electricity system. And different individuals and institutions have different perceptions of what specifically a "smarter" grid looks like and what it should do. A common theme across different definitions of smart grid is the further integration of information technology into electricity system management. As such, smart grid includes both hardware and software. It includes a variety of interlinked technologies including advanced meters and sensors, the management of "big data," and other technological configurations that enable increased reliability, more renewable electricity, and improved efficiency, resilience, and flexibility.

The many motivations for smart grid also include the potential to lower the cost of the system through efficiency improvements and managing peak demand. To produce electricity during periods of peak demand, utilities run expensive and inefficient plants, making the electricity more costly; if the demand for electricity during peak hours was reduced, fewer power plants would have to be built or maintained to meet these infrequent high-demand periods. A "smarter" grid could also promote more engaged electricity consumers, supporting both those who install their own renewable generation and those who more actively manage their electricity use. Consumers can track energy use through metering and make electricity use decisions that could save money through dynamic pricing that aligns the time-of-day price of generating

As long as cost matters more than environment, we will stay in the capitalist paradigm.

electricity to its use. An additional critical motivation is lower carbon and environmental emissions achieved through incorporating renewable generation and more efficient use of electricity. Some visions of smart grid also transform the oil-dependent transportation sector into another component of the electric sector with the integration of electric vehicles. From a societal perspective, smart grid also allows individuals and communities to ask new questions of the electric system. Beyond system reliability and affordable cost, smart grid has potential to spur social and behavioral change, including empowering individuals and communities to have more *- But not inevitable* localized control of and engagement in their energy choices. But, like change in all complex systems, smart grid also poses multiple potential downsides, ranging from increasing rather than decreasing costs and emissions to heightened concerns about cybersecurity, privacy, and health.

Acknowledging this intriguing landscape, in this book we explore both the social and technological complexities of electricity system change. Recognizing that electricity systems are composed of interlinked technologies, social practices, people, and organizations, we investigate these relationships. In the first half of the book we explore the heterogeneity of smart grid by describing variation in its promises and pitfalls, its technological configurations, and the actors and organizations involved in and influencing how smarter electricity systems are developing. The second half of the book moves beyond these heterogeneities to compare specific aspects of smart grid development: deployment of smart meters, integration of large-scale wind power in the electric system, community-based small-scale grid innovations, and connections with climate change. These latter chapters contextualize smart grid development through the exploration of geographic and social heterogeneity in different places by focusing on the struggles surrounding who has control and who benefits. By focusing on control, we are interested in the dynamics of who, where, when, and how different system actors are able to shape the electricity system. By focusing on perceived benefits of electricity system change, we are interested in who, where, when, and how different system actors benefit from smart grid. We explore the dynamic evolution of grid innovation, and we connect these changes with larger societal issues.

We focus on the social dimensions of smart grid, and also explore how they interact with the technological challenges. Our attention to the interactions between social and technical developments is intentional; this book is not a conventional engineering or technology text. Several other recent books and countless articles focus on the technologies of smart grid, and in Chapter 3 we will review the major categories of technologies that are often included under the smart grid umbrella. Rather than focusing primarily on the technological or engineering challenges, our approach to understanding smart grid is to explore the coevolution of social and technical systems and explore how they influence one another in expected and unanticipated ways.

1.3 Who Are We?

We are writing this book on smart grid and electricity system transformation because over the past decade we have become increasingly aware of the cultural and political embeddedness of energy system change. We are three professors who use interdisciplinary approaches to research the complexities of energy systems, environmental science, and engineering, policy, law, and environmental communication. We are motivated by a fundamental interest in the interconnected links between energy systems, societal change, and the environmental challenges associated with energy. Within the research community studying climate change and energy, a technological focus and accompanying economic logic dominate much of the research and associated policy discussions. But our perspective is different. Throughout the past decade our work has highlighted interconnected and embedded energy and climate challenges. Each of us has become increasingly aware of the often unrecognized social and political influences shaping energy systems. This book is an attempt to integrate and consolidate these less well-explored, yet critical, dimensions of energy system change.

Our work on energy systems reflects our own varied regional experiences. The three of us each work in different universities located in different parts of the United States and Europe, so we are experiencing electricity system changes in different contexts within our communities in New England (Vermont and central Massachusetts), in the Twin Cities area in Minnesota, and in central Texas and Sweden. We each have also lived at different times in our lives in different countries (including Australia, Belgium, Burundi, China, Ireland, Kenya, Sweden, and Tanzania). So our integrated perspective presented in this book is the culmination of our collective experiences and backgrounds and our interest in energy as a critical global concern.

In addition to being researchers and educators, all three of us are also mothers, sisters, daughters, and partners. We care about electricity systems and the potential changes that smart grid offers because we are deeply concerned about the future of the world our children and grandchildren will inherit. Each of us has a strong passion for understanding energy and environmental concerns, but we also have found a collective passion to better understand the dynamics of change and wrestle with why energy system change is so difficult. Our three-way collaboration has grown over the past decade in a way that has strengthened both our individual passion for understanding and revealing the complexities of energy system change and our humility in facing the magnitude of change. Together we challenge one another to broaden our individual tendencies to focus on particular aspects of the elephant. Our long-term collaboration has forced us to talk, think, and write in synergistic ways that result in more comprehensive reflections than any of one of us could produce by herself.

We began writing this book while staying in a net zero-energy passive-house renovated Irish cottage on the rugged and remote northwestern coast of Donegal in the summer of 2013. This house, which is the home of Jennie Stephens' parents, Cathal and Sarah Stephens, was recently renovated by Cathal Stephens, an architect with a keen interest in demonstrating the possibilities of technical and social change toward making buildings more sustainable (Stephens 2011). The home is powered by its own 6.6 kW wind turbine that takes advantage of the steady, strong winds of coastal Donegal. During a weeklong writing retreat that we spent at this house, we solidified our appreciation for the powerful feeling associated with the reliable and clean independence of working and living in a house that generates its own electricity. As we spent hours researching, writing, organizing our ideas, and structuring this book, we drank tea (ate chocolate) and gazed at the shifting colors of Trawenagh Bay. The window framed both the small turbine that powered the house and, in the distance across the Bay, several large industrial-scale wind turbines that spun steadily. Within this setting we were aware of the rapidly changing roles of energy and the growing role of wind resources in Ireland and around the world.

1.4 Emerging Tensions and Power Struggles

Stories of smart grid development are useful for studying and understanding energy system change because of the multiple tensions that emerge among the different actors' visions and interpretations of smart grid's promises and pitfalls. Power struggles are emerging in multiple complex ways in different regions, with different consequences. While the dominant paradigm of smart grid remains one of technological progress and utilitarian efficiency, smart grid development also highlights multiple emerging and entrenched struggles. Some of these struggles and tensions are demonstrated in the September 2013 release of the independent film *Take Back Your Power*, in which multiple skeptical views about why smart grid is being pushed, who may benefit, and who may pay the price are communicated in an investigative journalistic style. Our book, in contrast, is based on two multiyear, National Science Foundation-funded research projects in which we have analyzed hundreds of documents and interviewed dozens of individuals from many institutions who are involved and engaged in energy systems. Our research has led us to consider big and small questions, ongoing and emerging challenges, and the multiple tensions, coalitions, and inherent power struggles in creating the future electricity system. These tensions include incumbents versus new actors, perceived costs versus benefits, and questions of who pays, who plays, and who writes the rules. Questions of the timescales and the spatial scales over which costs and benefits are to be distributed come to the fore. Another set of tensions relates to actors' perceptions on whether smart grid should be oriented toward promoting a more centralized or a more decentralized electricity system, and/or whether both centralization and decentralization can and should be promoted simultaneously.

Good Q.

Another key tension in smart grid development relates to whether smart grid technology empowers consumers with more autonomy and control to manage their energy systems, or whether smart grid changes could result in disempowerment through a loss of privacy and control by individual households and electricity consumers. At this point it seems like smart grid could contribute to either and both – a more centralized electricity system and/or a more decentralized system; more opportunities for customer involvement in the energy system, or less. Other key tensions include whether a smarter grid would provide enhanced security to our current system or create new vulnerabilities for potential cyberattacks on the grid, and whether a smarter grid will accelerate a reduction in fossil fuel reliance by facilitating more renewables or whether fossil fuels will remain dominant and influence smart grid development in such a way that smart grid investments contribute to perpetuating fossil fuel dependence. Of course, the dichotomies are neither straightforward nor clearcut, as the case studies included in this book illustrate. Smart grid could enable increased renewables *and* an increase in coal use, as is the current situation in Germany. Smart grid could allow for distributed generation to enhance system reliability through the creation of microgrids *and* unintentionally exacerbate local air pollution. Smart meters and dynamic pricing could lower consumer costs for ten months out of the year and create a public (and political) revolt when high prices are passed on to unwilling electricity customers in an effort to link the price of electricity production and consumption during the peak summer months. We find the study of smart grid so interesting because the circumstances are rarely black and white, but rather marbled, shaded, and embedded within specific contexts.

An emerging struggle in many regions relates to how and to what degree solar PV owners should pay to maintain the distribution network. Homes and businesses with their own onsite solar electricity generation still rely on and benefit from being connected to the larger grid. When the sun does not shine these systems do not produce power. Unless the owners have invested in battery storage, they are dependent on purchased electricity for the hours in which the sun does not shine. They are also dependent on the larger grid to take any excess generated power that they do not use on-site. But how and to what degree these customers should support grid services when they are not purchasing much electricity from the utility remains unresolved. As solar PV produces power during the middle of the day, its value is often quite high. The challenge electric utilities face from high levels of solar and other forms of distributed generation is often (and dramatically) termed "the utility death spiral," as it undercuts utilities' current basic business model.

Whether or not smart grid is a useful term remains open for debate. While its widespread use in the past decade suggests that many seem to find it a convenient label to describe general electric system change, some people scrupulously avoid the term. We have already mentioned how it is an ill-defined, ambiguous, umbrella term that means something different to different people. Many have asked: is such a vague term useful? A 2011 MIT report entitled "The Future of the Electric Grid" intentionally avoided the label smart grid (MIT 2011). The authors of this report explained explicitly that they refrained from using this phrase because of its ambiguity. Other technical authors have also balked at its ambiguity and meaninglessness. Within the power sector, there seems to be a general shift toward the less polarizing term "grid modernization." However, smart grid retains its cachet.

These unresolved tensions and emerging power struggles result from a complex landscape of competing priorities and concerns. In this book, we explore this complexity by telling multiple stories about smart grid development in different places and across different scales. In these narratives, we demonstrate how individuals' perceptions of smart grid depend on their worldview and the priorities established within their cultural and professional spheres. Different actors support different dimensions of smart grid development, and see smart grid as fulfilling different societal goals. Some actors, particularly those who are skeptical and unsupportive of smart grid development, see smart grid as increasing risks associated with big government and corporate control in society, raising negative health and safety issues from smart meter radiation, reducing privacy from data energy consumption data, and enhancing the vulnerability of the grid to cyber-sabotage.

1.5 Our Approach

Electricity systems are an increasingly critical complex infrastructure that most people do not know much about. One goal of this book is to reveal and explain some of this complexity. A secondary goal relates back to the parable of the elephant and the blind men. Even among those who are well informed about the electricity system either through their professional work or their personal interest, it is clear that individuals and organizations (and even individuals within an organization) have very different perspectives, priorities, and understanding about the electricity system and its potential for change. So another goal of this book is to shine light on different aspects of the electricity system by exposing to all the breadth of smart grid visions, priorities, and perspectives. Perhaps with additional insights and understanding about others' perspectives, some of the tensions and struggles can be reduced. With enhanced mutual understanding made possible by the broad perspectives and narratives within this book, greater alignment of interests and priorities may evolve in way that accelerate positive system change.

For both experts and non-experts alike, understanding smart grid and the potential for electricity system change is based on their particular background and cultural, professional, and political values. Our goal in writing this book is to tell the story of smart grid from multiple different perspectives in such a way that any reader, whether new to the area or an experienced electricity system professional, will learn and gain understanding of the larger smart grid landscape. Given the critical importance and large scope and scale of electricity systems in our world, it is difficult to understand the whole system and all of the different perspectives within the system. This book attempts to shine light on multiple perspectives with an ultimate goal of helping different actors understand each other's priorities. We are writing this book to unpack and make sense of some of this complexity.

The opportunities and challenges of smart grid development vary significantly across countries and within regions of a single country. Electricity system change includes complex jurisdictional challenges. While this book incorporates mention of smart grid priorities and challenges throughout the world, many of the stories will focus on the United States, both because the U.S. context is the geographic and political area of the world in which we have the most experience, as citizens, electricity consumers, and researchers, but also because we seek to highlight the regional heterogeneity in smart grid development within the United States and illuminate the rich debates and discussions occurring across multiple contexts. We also draw on examples from Canada and Europe, and recognize that issues associated with grid development extend beyond these two continents.

An important perspective that we bring to our review and analysis of smart grid is that of sociotechnical systems. Central to this perspective is the notion that large technical systems coevolve with associated social, cultural, and political institutions. The trajectory of all technological change is intricately linked to social factors, and the trajectory of social change is intricately linked to technological factors. Constant, dynamic interactions among social and technological dimensions shape an interconnected complex system. This sociotechnical systems perspective has roots in sociological (Bijker, Hughes, and Pinch 1987) and historical (Hughes 1983) accounts of technological change, as well as in evolutionary economics and other influences.

Sociotechnical systems include technology, infrastructure, maintenance networks, and supply networks, as well as regulations, markets, user practices, and cultural meaning. Sociotechnical systems can become quite stable and resistant to change when the social and technical dynamics form reinforcing mechanisms to protect and promote the entrenched regime (Turnheim and Geels 2013). The status quo is perpetuated and strengthened as established actors, institutions, and technologies contribute to maintaining current arrangements. But a sociotechnical system can become unstable when there is alignment of pressures pushing toward system-wide change (Geels 2005). When this happens the system can transition to a novel configuration that could eventually stabilize as a different system.

In this book we take a systems approach to move beyond the conventional linear view of science, technology, and innovation that assumes scientific research leads to technology advancements which lead to innovations (Keller 2008, Luhmann 1989). We embrace a broader view that incorporates the social dimensions of system change and acknowledges inevitable negative social and environmental consequences of technological development. We integrate our varied backgrounds and experiences to move beyond the technical and economic perspectives of electricity system change to expand energy system consideration to include key social, political, and cultural dimensions.

We embrace the notion that sociotechnical systems are dynamic and ever-changing, and that some sociotechnical systems are more stable than others and, therefore, more resistant to transitions than others. We also acknowledge that system change is extremely difficult because of the reinforcing power of incumbent actors and institutions who often cling desperately to the status quo (Breslau 2013, Laird 2013).

1.6 Organization of the Book

This book takes readers on a guided tour of the social and technical complexity of smart grid. We describe the overarching social context of smart grid and how it is changing over time. With this book, we attempt to make visible a topic and a critical societal infrastructure that is often invisible, or at least overlooked.

We begin with three chapters that lay out the basics of this smart grid map. Chapter 2 presents the dominant promises and pitfalls that are most often associated with smart grid. This chapter describes a broad spectrum of perceptions, including the views of technological optimists who think smart grid has potential to solve many of humanity's most vexing problems as well as the perspective of mistrustful skeptics who see smart grid as an expansion of corporate control over individuals' lives. Chapter 3 then explains the different technological components that are most often considered to be critical pieces of the smart grid puzzle. This chapter provides background to understand technological configurations that are often included in smart grid discussions. Chapter 4 then reviews the key actors involved in smart grid development and deployment, explaining who is involved and how different actors are engaged, and exploring their dominant priorities.

Chapters 5, 6, and 7 focus on three particularly important aspects of smart grid development: smart meters, integration of large-scale wind power, and the development of small-scale, community-based electricity systems. The stories within these chapters contextualize the complicated intersection of promises, pitfalls, technologies, and actors in different settings. The geographic and social heterogeneity of smart grid development is demonstrated in these chapters as we focus on two themes: who has control and who benefits. In addition to exploring who, we explore when, where, and how control and benefits are realized and perceived.

Chapter 5 focuses on smart meter installation efforts and the associated controversies and struggles. Smart meters are the smart grid technology that most directly interfaces with consumers. Smart meters which are coupled with time-of-use pricing offer the promise of aligning price incentives with system costs to allow individual households to better manage and control their energy use and allow them to benefit by saving energy and money. However, some customers are suspicious of and concerned about the meters, and public opposition has powerfully altered smart meter deployments in some communities. Divergent actors' priorities and perspectives on smart meters are illustrated in this chapter by exploring the rollout of smart meters in two locations in the United States (California and Massachusetts), and in one European country, Germany.

Chapter 6 then tells of the interlinked development of large-scale wind power and smart grid. In this chapter, we explore the growth and integration of wind power in the electric power system and explore the pivotal role smarter grids play in enabling large-scale renewables and governing the creation of new sociotechnical systems. Wind power development has been shaped by its sociotechnical context and regional determinants, so here we focus on wind power development and its dynamic interactions with policy, the grid, and energy markets. We develop three in-depth case studies: 1) Texas, the U.S. state with the most installed wind power; 2) the Upper Midwest of the United States, where states and the electric grid system operator have worked together to plan for and integrate wind power into the electric system; and 3) Germany, a nation in the midst of an energy transition, the "*Energiewende*," and a leader in the development of large-scale wind power. Together, these cases allow us to more deeply explore how a smarter grid is enabling both technological advances as well as shifts in laws and markets and associated regulatory, financial, and legislative institutions.

Chapter 7 explores a growing movement to support small-scale distributed generation and community-based energy initiatives and systems, empowering local control of electricity systems. Within this chapter we explore individuals, communities, and organizations taking control of their own local electricity systems. We introduce and describe the notions of community-based energy, microgrid, nanogrid, locavolt, and prosumer to explore small-scale energy systems and local control and benefits. This chapter uses cases to demonstrate how different key actors are attempting to harness technologies to achieve the promises of smart grid for local benefits. We highlight several small-scale smart grid initiatives, identify the technologies, and finally identify actors and interests most directly impacted by small-scale smart grids. Our first case tells the story of how the city of Boulder, Colorado is attempting to municipalize its electricity distribution system, separating itself from its current investor-owned utility which has been providing the city with electricity since 1928. Our second case is the Pecan Street Project in Austin, Texas, which is an example of a well-funded smart grid pilot project with a specific neighborhood focus. The third case focuses on the efforts of the island of Bornholm, Denmark in building an independent electricity system or microgrid. In this case, we see that decentralization has been driven by reliability concerns and encouraged by centralized EU and Danish national-level policies. After developing the cases, we present a few additional interesting examples of small-scale community initiatives and then summarize the commonalities and differences across contexts, and explore possibilities for integrating these small systems into the overall vision of smart grid.

Chapter 8 focuses on the challenges of confronting climate change and explores the tensions associated with how smart grid could contribute to climate mitigation as well as climate adaptation, including resilience and preparedness. In addition to reviewing common assumptions about the roles a smarter grid could play in a changing climate, this chapter explores the more provocative and critical possibility that a future smart grid electricity system could exacerbate, rather than reduce, climate risks. Given the increasing importance of considering climate change in all discussions of energy, this chapter provides a valuable perspective on an additional set of tensions regarding the environmental and climatic impacts and potential of a smarter grid.

Chapter 9 concludes with a call for broadening smart grid conversations to advance collaborative thinking and engagement on the social implications of energy system change. We highlight the importance of a sociotechnical perspective that elevates social considerations and moves beyond the dominant and narrow technical and economic perspectives on smart grid. Acknowledging the diversity of smart grid struggles, smart grid meanings, and smart grid opportunities in different contexts, we return to considering the ways in which smart grid offers evolutionary change, revolutionary change, both, or neither.

We hope that readers of this book will gain insights and perspective on the size and complexity of the smart grid elephant. Although we are aware of our own limitations in terms of providing a comprehensive description of every perspective of the electricity system, we do represent multiple holistic views that integrate critical cultural and social dimensions which are often overlooked in the reductionist techno-economic perspectives which dominate descriptions of energy systems.

References

ASCE. (2013) American Society of Civil Engineers 2013 Report Card for America's Infrastructure. www.infrastructurereportcard.org/a/#p/home.

Bijker, W. E., T. P. Hughes, and T. Pinch. (1987) *The Social Construction of Technological Systems*. Cambridge, MA and London: MIT Press.

Breslau, D. (2013) Studying and Doing Energy Transition. *Nature and Culture*, 8.

Geels, F. W. (2005) The Dynamics of Transition in Socio-technical Systems: A Multilevel Analysis of the Transition Pathway from Horse-Drawn Carriage to Automobiles (1860–1930). *Technology Analysis and Strategic Management*, 17, 445–476.

Hughes, T. P. (1983) *Networks of Power: Electrification in Western Society 1880–1930*. Baltimore, MD: Johns Hopkins University Press.

Keller, K. H. (2008) From Here to There in Information Technology: The Complexities of Innovation. *American Behavioral Scientist*, 52, 97.

Laird, F. N. (2013) Against Transitions? Uncovering Conflicts in Changing Energy Systems. *Science as Culture*, 22, 149–156.

Luhmann, N. (1989) *Ecological Communication*. Chicago, IL: University of Chicago Press.

MIT. (2011) *The Future of the Electric Grid: An Interdisciplinary MIT Study.* Cambridge, MA: MIT.

Moniz, E. (2013) www.greentechmedia.com/articles/read/DOE-Head-Ernest-Moniz-Delivers-First-Major-Policy-Address

Stephens, C. (2011) Tradition, Character and Energy Efficiency: An Example of Sustainable Renovation/Expansion of a Traditional Irish Cottage. Paper presented at 27th *International Passive and Low Energy Architecture Conference*, Louvain-la-Neuve, Belgium, July 13–15, plea-arch.org/ARCHIVE/2011/2011_Proceedings_Vol2.pdf.

Turnheim, B. and F. W. Geels (2013) The Destabilization of Existing Regimes: Confronting a Multi-dimensional Framework with a Case Study of the British Coal Industry (1913–1967). *Research Policy*, 42, 1749–1767.

Webley, K. (2012, November 26) Hurricane Sandy By the Numbers: A Superstorm's Statistics, One Month Later. *Time*. http://nation.time.com/2012/11/26/hurricane-sandy-one-month-later/

2

Promises and Pitfalls of Smart Grid

2.1 Changing Expectations of Electricity Systems

Access to electricity offers tremendous benefits for society and individuals. As mentioned in Chapter 1, our dependence on electricity has increased dramatically, and we now rely on electricity for fundamental communication, food, health, transportation, and other needs. As we have become more dependent on electricity, we have also become more aware of the negative societal impacts of our current fossil-fuel combustion-dominated approach to producing electricity.

The electricity system is expected to provide continuous access to reliable and affordable energy. Reliability and affordability have stood out as the prime system directives: system engineers and economists have developed and optimized the system based upon these goals. Both political and regulatory constructs use reliability and affordability as guiding principles shaping the electricity system. But we now find ourselves in an era where additional expectations are being added to the reliability and affordability directives.

These other expectations are growing and becoming simultaneously more diverse and less predictable. Electricity systems are increasingly expected to be prepared for more frequent and intense storms, to rapidly respond to any disruptions, and to minimize all kinds of environmental impacts of their operations. The environmental impacts of fossil fuel combustion include emissions of sulfur dioxide, oxides of nitrogen, mercury, and particulate matter that degrade air and water quality and harm human and ecosystem health. The electric sector emits 40 percent of U.S. greenhouse gases (GHG). Fossil fuel use also causes ecological degradation during extraction; coal mining, natural gas fracking, and oil drilling are all associated with environmental as well as social impacts influencing communities. Other sources of electricity, such as nuclear, large-scale hydropower, and wind projects also affect the environment. Increased awareness of the negative societal and environmental impacts of electricity generation is driving new demands and expectations of electricity systems. The notion of smart grid has emerged in response to these new demands and expectations.

Implementing changes to electricity systems to meet these many different expectations poses multiple challenges due to uncertainty and diversity of opinion on how to prioritize. Smart grid offers multiple promises, but not all smart grid initiatives are able to simultaneously meet all expectations. Priorities for electricity system change span a broad spectrum, with different individuals and organizations having disparate expectations and perceptions of what is possible and what changes should be made. This broad range of priorities creates inevitable tensions that are now being felt. While environmental quality is fundamentally important for some smart grid supporters, it is peripheral for others. The potential for smart grid to empower people to become more engaged and involved in their electricity systems captures a set of socially oriented promises that are very appealing to some; on the other hand, these assumptions for cultural change are disregarded by others as unrealistic, impractical, or unnecessary.

Maintaining a reliable and secure electricity system is considered critical for economic and political stability, which means that all changes in the electricity system potentially have both economic and political implications. Smart grid, therefore, is associated with a broad array of promises as well as pitfalls.

While the primary goal of many involved in developing future electricity systems is to provide low-cost, reliable access to electricity, some have also embraced smart grid as a technological platform that promises to address larger societal ills. Perceptions of the potential of smart grid span a broad range with a utopian optimism at one extreme and a dystopic pessimism at the other, with most individuals and organizations falling somewhere in between.

At the optimistic end of the spectrum, smart grid promises a new electricity world where revamped systems provide perfect alignment in improving multiple social as well as environmental challenges and contributing positively to the human condition. In this utopic smart grid vision, increased automation of the grid allows for a reliable "self-healing" infrastructure which integrates renewable energy production, seamlessly balances energy supply and demand, and allows for citizens to drive clean and quiet electric vehicles. The negative environmental and health impacts of electricity production would be eliminated by the more sophisticated electricity system. Citizens would breathe crystal-clear air, as emissions of air pollutants like SO_x, NO_x, and particulates that contribute to childhood asthma (EPA 2013a), and greenhouse gas emissions that contribute to anthropogenic climate change (EPA 2013b), would be further reduced by the transition away from fossil fuels to renewable-based generation. This optimistic perspective is captured in the following quote from the Natural Resources Defense Council: "The smart grid can give us cleaner air, better health, lower electricity bills, and reduced carbon dioxide (CO_2) emissions in the atmosphere" (NRDC 2012).

On the pessimistic end of the spectrum, smart grid includes pitfalls that pose unacceptable risks to both individuals and society, including high economic costs, increased social inequality, worsened environmental and health impacts, and diminished cybersecurity and individual privacy. A recent article in *Counterpunch*, a well regarded investigative journalism publication, demonstrates these extreme views by describing smart grid as "an eco-health-safety-finance debacle with the potential to increase energy consumption, endanger the environment, harm public health, diminish privacy, make the national utility grid more insecure, cause job losses, and make energy markets more speculative" (Levitt 2011). This negative perspective evokes concerns about "Big Brother" and underscores fears of the expanding reach of corporate and/or government entities infringing on the lives of individuals and communities.

Beyond the extremes at either end of the spectrum, most people see both promises and pitfalls of smart grid. Generally, most people emphasize the positive potential of smart grid more than its negatives. Our recent research on media analysis of smart grid found that newspaper articles in the *New York Times*, *Wall Street Journal*, and *U.S.A. Today* all report the positive attributes of smart grid potential more than negative aspects of the technologies, with a ratio of four positive mentions to every negative. In our focus groups and interviews with more than 200 stakeholders professionally involved in the electricity system, we also found that participants were more likely to introduce and discuss the promises, rather than the pitfalls, of smart grid. Deployment of smart grid – particularly installation of smart meters – has, however, engendered intense discussion of the smart grid pitfalls. Recent social science research on smart grid in the Canadian provinces of British Colombia, Ontario, and Quebec has shown that the pitfalls of smart grid are especially prevalent in the public discourse during periods of smart meter deployment, but once the meter deployment phase is finished, the promises once again become more prominent (Mallett et al. 2014).

Although we recognize the existence of exclusively technological promises and pitfalls, our focus throughout this book is on the dynamic sociotechnical interactions between technical systems and larger societal contexts. These interactions are especially relevant for legitimacy in democratic systems. So in this chapter we focus on the broad social and technological categories of promises and pitfalls as they relate to:

(1) reliability and security;
(2) the economy;
(3) environmental quality;
(4) citizen empowerment.

While some of these promises and pitfalls are relatively independent of each other, others are interconnected in multiple, complex ways. The goal of this chapter is to highlight the spectrum of perspectives, including both the promises and the pitfalls, of smart grid. To set the stage for the rest of the book, this chapter presents the broad range of positive and negative perceptions of smart grid prioritized by different actors across different contexts. The chapter is structured to first present the positives, then the negatives, and then concludes by highlighting the policy implications of the tensions between the two extremes.

We explore the ambiguous and interconnected nature of smart grid, focusing on the most vociferously argued smart grid "imaginaries" (Jasanoff 2006). We realize that, as with most public controversies, those who focus on either the extreme positive or the extreme negative are often dismissed; the prophets of utopian and dystopian futures are often vigorously decried as noisy extremists. In the case of smart grid, however, these voices are directly influencing smart grid implementation. The extreme as well as the more mundane perspectives are critically important to understand because they set the stage for public conversations about sociotechnical change in electricity systems.

This chapter provides a foundation for subsequent chapters exploring the breadth of different technologies (Chapter 3) and stakeholders (Chapter 4) associated with smart grid development, and the more detailed discussions of emerging tensions in smart grid development presented in the second part of this book (Chapters 5, 6, 7, and 8).

2.2 Promises of Smart Grid

Smart grid promises an improved electricity system with multiple benefits; many of these benefits are represented in the idealized vision of a smart grid future in Figure 2.1. The many promises of smart grid include a more reliable, resilient, and secure energy sector; a stronger economy; a cleaner environment; and a more

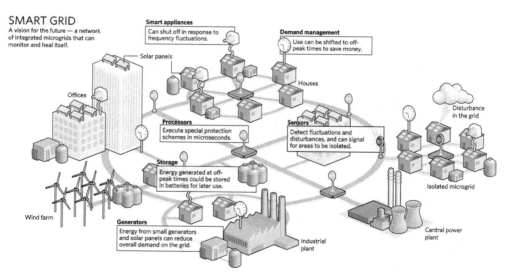

Figure 2.1 A representation of an idealized vision of a smart grid future. Reproduced with permission from Nature Publishing. Source: Maris, 2008.

empowered and engaged citizenry. In this section we describe the dominant promises often put forward to justify investment in moving toward and investing in electricity system change. While the multiple promises of smart grid can be categorized into distinct types of societal benefits (reliability, environmental improvement, efficiency of resources, etc.), most of these are intricately linked with each other. The claim that "smart grid would help make everything better" (Kowalenko 2010) summarizes a utopian vision of smart grid. In this optimistic vision, electricity systems "should deliver more power, more reliably, and with greater efficiency, wherever and when-ever needed. Outages and brownouts should be infrequent, localized, and quickly resolved. Less energy should be lost in generating, transmitting, and delivering electricity, and every conceivable source of electric power should be used" (Berger 2008). Researchers at Lawrence Berkeley National Laboratory in the United States add that smart microgrids show "great promise for bringing basic electricity services to people who currently lack them" (Nordman 2010).

One of the most basic smart grid promises is increased reliability, resilience, and security achieved by a sophisticated information communication technology overlay of networked sensors, big data, and automated interconnections across the system (Amin and Wollenberg 2005). Investments in smart grid also promise opportunities for economic growth for multiple constituents, including individual consumers, electricity providers, and society more broadly (Amin 2013a). Smart grid also promises environ-mental benefits, both through integration of renewable resources and by reducing demand by encouraging electricity conservation through changing patterns or electri-city use. Smart grid also promises to enable and empower consumers to participate in decisions about how the electricity system is envisioned and deployed (Amin 2013b). The smart grid future envisioned by many smart grid proponents promises to make everyone's life better through the realization of one or more of these promises.

2.2.1 Enhanced Reliability and Security

For many, the most valuable promise of smart grid is enhanced reliability and security of the electricity system. This includes both system operation and planning: smart grid could allow increased resilience to and preparedness for disruptions as well as reduced vulnerability to disruptions. Disruptions to electricity systems can be caused by physical disturbances related to weather or natural disasters as well as by social and political disruptions related to the geopolitics of fuel or equipment supply.

Improved Reliability and Resilience

The increased ability of electricity systems to respond to both natural and human-caused disruption can increase system reliability and resilience by improving "wide-area situational awareness," which is a term used in the electricity industry to represent the capacity to monitor and be aware of what is going on throughout the entire system. A key component of improved reliability includes communication sensors that would make it easier to identify where and when a disruption occurs. Examples of improvements to resilience include networks of sensors on the high-voltage transmission grid and low-voltage distribution networks that can allow for better monitoring and management of the system. On the high-voltage transmission network, synchrophasors provide real-time measurements of electricity waves to monitor frequency and rapidly identify system disturbances. On distribution net-works, sensors and advanced communication devices allow better management of power quality. An additional component of reliability and resilience relates to adding and integrating distributed generation (DG) on the distribution network and creating islanding capabilities and microgrid management to improve overall performance and reduce risks of major cascading outages (Kowalenko 2010).

A key part of smart grid's promise of enhanced reliability through improved resilience relates to the idea of creating a "self-healing" system, allowing for automated fault detection followed by automated repair. Continuous system-wide monitoring and local islanding, or isolating a subsection of the grid during times of disruption, could enable quicker recovery from storms or other power loss events and make the electri-city system more robust in times of system duress (LaMonica 2012). For example, during Superstorm Sandy in 2012, Princeton University was able to keep the lights on by disconnecting from the bulk power grid, creating its own island of power, its campus-system microgrid. More details of this example are given in Chapter 7.

Both civilian and military interests are enthusiastic about the resilience promised by smart grid. After determining that the current electrical grid poses an unacceptably high risk of power outages, the U.S. Department of Defense is cooperating with the U.S. Departments of Energy and Homeland Security to develop a program called Smart Power Infrastructure Demonstration for Energy Reliability and Security (SPIDERS), promoting microgrids that would enhance system resilience by ensuring "continuity of mission-critical loads" in the event of widespread and/or long lasting power outages (Perera 2012; Sandia 2012).

These examples demonstrate the power of the smart grid promise of improved reliability and system resilience. The potential of rapid fault detection for building self-healing capacities through deployment of sensors and automation of advanced monitoring components has great appeal. Chapter 3 provides more details on the specific technological components that relate to enhancing reliability and resilience. Chapter 4 identifies the actors most closely associated with increasing system resili-ence and reliability.

Improved Cybersecurity

Smart grid's promise of enhancing reliability is linked with multiple promises related to improving security, particularly cybersecurity. Some smart grid enthusiasts make the case that a smarter grid will enhance cybersecurity and therefore enhance the reliability and security of electricity systems (Kurada, Dhanjal, and Venkatesh 2013). As will be discussed in section 2.3 on pitfalls, some actors, in contrast, see cybersecurity as a significant smart grid weakness due to the system's vulnerabilities to hackers or openings to outside malware.

Some smart grid proponents see increased monitoring as a critical promise which would increase system operators' capacity to quickly detect abnormalities that stem from malicious attacks by differentiating between intentional and accidental anomalies. A smarter grid could provide operators with early warning signals when security is breached and allow them to identify the proximate cause of the breach. Some smart grid advocates point out that the extensively networked system excludes the option of sustaining a pre-internet electricity system. Although current electricity systems do not necessarily include smart technologies such as communication sensors that could improve security and strengthen reliability, current systems already rely on significant inputs that are only available through the internet (Kurada et al. 2013). This means that the current system is already vulnerable to cyberattack. One promise of smart grid is that it could protect against, and enable differentiation between, accidental faults and nefarious political threats.

Recognizing that cybersecurity has become a rather generic buzzword, here we use the Information Systems Audit and Control Association's (ISACA's) definition, which is "the sum of efforts invested in addressing cyber-risk, much of which was, until recently, considered so improbable that it hardly required our attention" (Barzilay 2013). Within this context, smart grid enthusiasts argue that the same electronic sensors and other intelligent components that promise increased system resilience also promise enhanced cybersecurity. This chapter provides additional detail on theft prevention in the section on economic promise, and further explores cybersecurity risks in the section on pitfalls.

Energy Independence for Improved Geopolitical Security

Another critical smart grid promise of enhanced security relates to increased energy independence. In the United States, Germany, Japan, and many other nations, the political awareness of a smarter grid is embedded in the perceived benefits of national energy self-sufficiency. The geopolitics of energy are complex, costly, and contentious. While most countries rely on global energy markets for their oil, natural gas, coal, uranium, or electricity, the economic and political vulnerabilities of energy dependence are increasingly evident. Smart grid is part of energy system reforms to enhance energy independence.

In the United States, for example, the Energy Independence and Security Act (EISA) of 2007 highlights the perceived value of achieving national energy independence (Congressional Research Service 2007; U.S. Department of Energy 2008; U.S. Government Printing Office 2008). In spite of increasing domestic oil production, the United States spends roughly $30 billion a month on oil imports. The electrification of the transportation sector could help to move the United States toward greater energy independence and security. The smart grid promise of energy independence extends beyond the United States. The Danish government has committed to achieving energy independence by 2050, and smart grid development is central to its strategy. Because Denmark has no native fossil fuel resources, energy independence requires linkages to a Nordic power grid and independence from fossil fuels. Lykke Friis, the Danish Minister of Climate and Energy, refers to the nation's energy strategy as "a declaration of energy independence" (Danish Government 2013). Chapter 7 provides more details on the operation of Bornholm's microgrid and its contribution to EU energy goals. In Germany, increased integration of local renewables could reduce dependence on Russian natural gas.

Smart grid promises to be part of a strategy that allows the United States to develop independence from the vagaries of political regimes in the traditional petroleum-producing nations. "American energy independence means freedom to produce our own electricity and freedom to sell it at fair market rates" (Hertzog 2013). Smart grid promoters point out that this independence does not stop at the national level, but can also be taken to the individual consumer. Collier explains, for example, that smart grid simultaneously enables U.S. energy independence at the national level and enables individual U.S. consumers to function independently of regional and national grids (Collier 2013). Steven Wade, an economist for the U.S. Energy Information Administration, explained that, as part of the shift wherein "we as a society are valuing energy independence more," smart grid is becoming increasingly attractive (Matthews 2013). Section 2.2.4 discusses the promises of consumer engagement and autonomy in more detail.

2.2.2 Strengthened Economic Conditions

By making the grid more efficient and reliable, smart grid also promises to strengthen the economy by delivering a host of economic benefits to many stakeholders, ranging from individual consumers at the household level to large industrial customers, utilities, and other electricity providers. Through better management of the system and its externalities, consumers could pay less for electricity, communities could reduce municipal energy expenditures and associated pollution, and states and countries could benefit from economic growth as a result of efficient, cleaner, low-cost electricity systems.

Smart grid promises these economic benefits in part by changing consumer behavior, enabling price-responsive demand-side management (DSM), and enhancing control and communication throughout the system (Charles River Associates 2005). By providing information on system costs to consumers, a smarter grid can help to align actual system costs with energy prices. Smart meters installed in homes and businesses can provide electricity customers with real-time data on their energy

? More productive than fossil fuels ??

use and costs. Chapter 3 includes more details on smart meters and other technologies that may contribute to a more economically productive energy sector, Chapter 4 describes the various actors who stand to benefit from these changes, and Chapter 5 provides a detailed case study on smart meter deployment.

Economic Benefits for Consumers

One of the promises of smart grid is lower electricity costs for consumers through mechanisms that allow them to reduce their electricity use and better match their use to dynamic price signals. Currently, most U.S. electricity customers pay a flat charge per-kilowatt-hour (kWh) and receive a monthly bill after the electricity has been consumed. Providing real-time price information through in-home or in-business displays could enable consumers to actively manage their energy use and shift their electricity use away from times when electricity costs are high, saving the entire system money. One way of doing this is through price-responsive DSM (Charles River Associates 2005), or dynamic pricing (EPRI 2011; Kowalenko 2010). Dynamic pricing offers price signals to consumers so they can respond actively to changing conditions by, for example, reducing electricity consumption when rates go up (Jessoe and Rapson 2013). Smart metering, real-time energy use information, and more detailed billing offer the potential to expand DSM to offer dynamic pricing to more customers.

There are different types of dynamic pricing programs, such as "time-of-use pricing," which may have different rate blocks throughout the day; "critical peak pricing," which charges high rates during emergency conditions such as a hot summer day; "variable peak pricing," where the time periods are set but the rates of the periods can vary with the market price; and "real-time pricing," where the retail rates reflect the market rates (Jones and Zappo 2014). Each of these different rate-pricing structures aims to better link consumers' rates to the actual cost of electricity. Although these pricing mechanisms have been widely promoted by economists and used by many industrial customers, they remain lightly used in the residential sector. While many small-scale experiments have used different pricing tools to model consumer behavior, many public regulatory commissions remain wary of approving dynamic price-rate structures. While states with restructured retail markets offering retail choice offer dynamic pricing to customers and some large utilities have residential pricing plans with high enrollments, most residential customers do not participate in this type of program.

Although the cost of producing electricity varies with the time of day, most residential electric customers in the United States and many other places still pay a fixed charge per kWh and receive a monthly bill from their utility. A smarter grid could modernize this system and give electricity consumers information on electricity generation costs. In this scenario, future households and businesses manage their electricity use and may save money by making more informed choices about electricity use. Smart grid could also automatically control appliances that can be programmed to respond to price signals. For example, an internet-connected refrigerator in a smart grid system could be programmed to cycle its compressor when electricity prices are lowest, or be remotely controlled by the utility to help reduce peak demand.

A smart grid could also offer reduced systemwide costs. These reductions in system cost could come directly from efficiency improvements, from better management of system externalities including environmental losses (discussed in Section 2.1.3 on improved environmental quality), from self-healing (discussed in Section 2.2), or from lower generation costs due to better use of resources or reduced peak demand.

Economic Benefits for Utilities

Smart grid promises multiple economic benefits for incumbent utilities, including reducing labor costs, enabling more sophisticated demand management, improving billing accuracy, enhancing customer engagement, and allowing more efficient use of capital resources. Smart grid also has potential to provide economic benefits associated with deterring electricity theft, increasing returns for infrastructure investment, and reducing costs of recovery after disruptions. By linking electricity generation to electricity demand in novel ways, smart grid creates new market opportunities for responding to consumer demands.

When utilities are justifying smart grid investment, two areas stand out: increasing demand response and reducing costs (both system and labor costs). As mentioned in the previous section, smart grid is a promising platform for developing more demand response opportunities. Because utilities pay variable wholesale market rates but are reimbursed at flat retail rates, utility managers generally favor demand response programs because these allow them in the short-term to reduce high-cost peak demand and, in the longer term, to postpone construction of new generation facilities. Unlike energy efficiency programs, which directly undercut utility revenues by reducing electricity sales and may require special rate-reimbursement programs like decoupling to encourage compliance, demand response is often viewed as a win–win situation.

Smart grid can also help utilities to reduce labor costs. By deploying smart meters, utilities can eliminate manual meter-reading and remotely monitor power quality, customer connections, and customer disconnections. In a time of economic recession, these cost savings and associated job reductions have not always been viewed positively. For example, Quebec Hydro faced large-scale protests from union members when it attempted to eliminate meter-reading jobs.

Smart grid offers the promise of helping utilities to reduce electricity theft, or informal grid connections. In Italy, where it was estimated that 40 percent of electricity was stolen, theft reduction was the primary motivation for deploying smart grid technologies (Scott 2009). In many developing countries, more than half of the available electricity is stolen through informal connections. This electricity theft negatively affects the ability of those managing the electricity system to invest in and improve the power system. In the U.S., an estimated $6 billion of electricity is stolen each year through illegal connections (Kelly-Detwiler 2013). The tamper-detection feature in smart meters provides utilities with real-time alerts if a meter has been altered, allowing rapid response to thwart potential theft. Using these meters, electric utilities can remotely disconnect non-paying users, which offers significant cost-saving and improves worker-safety. The scale of technological change possible with smart grid offers multiple potential economic opportunities for the incumbent utilities.

Economic Benefits to Other Actors

Smart grid also provides economic opportunities to new entrants in the electricity sector. As new devices, technologies, and software are developed, smart grid investments offer billions of dollars in new investment into companies. Smart grid also strengthens the case for investing in research and development (R&D) in the electricity sector by offering the possibility of new markets, cost recovery, and new opportunities for companies that generate, transmit, and sell electricity and associated services (Nemet and Kammen 2007). This spans both investments in hardware, such as new smart meters and new operations, and investments in software and new management tools. Opportunities for new actors can be seen clearly with the increase in demand response programs and the growth of companies that facilitate changes in efficiency and electricity demand. These companies, sometimes referred to as third-party demand response aggregators, have become active market participants in some areas and demand response has become a valuable grid resource that influences overall management of the grid (ISO-NE 2013; PJM 2013).

Societal Economic Benefits from a Smarter Grid

The economic benefits of smart grid also include indirect benefits resulting from a strengthened, more robust and efficient infrastructure and internalizing environmental externalities, discussed in more detail later in the chapter. By reducing the frequency and duration of costly power outages and improving system efficiency, a smarter grid has the potential to provide diffuse economic benefits to all members of society. These benefits result directly and indirectly from lowering the societal cost of generating, transmitting, and distributing electricity and enhancing management of consumer demand, and ensuring a robust and reliable system with minimal disruptions. Moving toward a smart grid system requires significant investment, but initial estimates by the Electric Power Research Institute (EPRI) project an excellent rate of return, with a cost-to-benefit ratio of 4:5 (EPRI 2011). Other potential economic benefits for society include an expanded high-tech industry, a growing renewable energy sector, and higher quality power. With smart grid the United States has an opportunity to improve the robustness of the economy through investment in its aging energy infrastructure, and potentially raise the D+ grade received on the American Society of Civil Engineers' infrastructure report card (ASCE 2013).

Smart grid promises increased efficiency across the power system by providing ways to use capital-intensive resources more efficiently (EPRI 2008). To ensure system reliability, some power plants only run for a few hours per year. Despite the inefficiency, these power plants are maintained and kept in service because the system must be able to handle peak loads. Commentators quip that this is like sizing a church to handle Easter Sunday crowds, or building a mall parking structure large enough for all of the Black Friday customers to park. System operators know that demand varies during the day and seasonally, but the current power system does not allow them to manage demand as efficiently as they could. Development of a smart grid would let system operators monitor and manage demand and generation. Advances in smart metering could allow direct control of industrial, commercial, and residential customer load, not just in response to system emergencies, but as part of normal system operation. Chapter 5 explores more details of smart meters' potential to contribute to efficiency improvements.

2.2.3 Improved Environment Quality

For some stakeholders, the most compelling reason to invest in smart grid is its potential to improve the environment by reducing negative environmental impacts of the energy sector. By making the system more efficient and deploying low-carbon renewables, the electric system can reduce greenhouse gas emissions and other environmental impacts (Jones and Zoppo 2014).

Deploy More Renewable Resources

By enabling the deployment and integration of variable renewable resources such as wind power and solar PV, smart grid can reduce the environmental impacts of the electric power system. A U.S. study by EPRI estimates a smarter grid could allow large-scale integration of variable renewable resources such as wind and solar, which could help decarbonize energy systems (EPRI 2008). For example, the incremental impact of the wind integration supported by smart grid is estimated to enable avoiding an additional 18.7 to 37.4 million metric tons of CO_2 by 2030 (EPRI 2008). Smart grid technologies could also enable distributed generation such as solar PV. For some, increased renewable generation also includes development of additional electricity storage; for others it means additional high-voltage transmission lines. Electricity storage and increased transmission lines could also enhance systemwide efficiency and reduce total generation needs. Integration of more renewable resources into energy mix promises a suite of environmental benefits, including lower air pollutant emissions, reduced water pollution, and lower CO_2 emissions. Chapter 6 explores the connections between smart grid and renewables in more detail by focusing on the coevolution of smart grid and the development of large-scale wind power.

Contribution to Climate Change Mitigation and Adaptation

Smart grid offers multiple promises related to responding to climate change. Electricity generation relies on fossil fuel combustion and emits 26 percent of global greenhouse gas emissions and 41 percent of all CO_2 emissions (IEA 2012). Although the Intergovernmental Panel on Climate Change (IPCC) stated that an 80 percent reduction in greenhouse gas emissions by 2050 is required to stabilize atmospheric levels of CO_2 (IPCC 2007), electricity generation is projected to grow 70 percent by 2035 (EPA 2010; OECD 2012; Outlook 2012). One of the most compelling reasons for integrating renewable energy sources into the electricity system is the potential they offer to help mitigate climate change. By increasing the proportion of electricity produced from renewable sources such as wind and solar, smart grid can help to lower the carbon intensity of the electricity system without cutting service or power quality (Hoag 2011).

In terms of climate adaptation, smart grid allows for enhanced system planning and the resilience provided by continuous system-wide monitoring. Microgrids also provide the ability to enable local islanding to protect electricity reliability during disruptions on the main grid. Together, these technical improvements can help system operators better adapt to extreme and variable weather events. Building a grid that is more resilient to climate-fueled disruptions is a critically important promise. Chapter 8 explores in more detail the potential of smart grid to contribute to both climate mitigation and climate adaptation.

Electrification of Transportation

One of the grand promises of smart grid is its promise to enable the electrification of the transport sector. In the United States, transportation contributes ~30 percent of greenhouse gas emissions, 22 percent of methane (CH_4) emissions, and 46 percent of nitrous oxide (N_2O) emissions (EPA 2013c). The electrification of transport promises to significantly decarbonize the energy sector (Tran et al. 2012), and could also encourage indirect improvements in environmental quality as plug-in hybrid electric vehicles (PHEVs) displace light trucks, SUVs, and vans (Peterson, Whitacre, and Apt 2011). Transport sector electrification could also enhance energy security, because most of the energy consumed for transportation comes from oil. A smart grid allowing for greater penetration of electric vehicles could potentially reduce these emissions by coordinating charging and electric system operations.

There are, of course, numerous scenarios regarding the promised benefits of integrating plug-in electric vehicles (PEVs) into the electricity system. For example, models integrating plug-in hybrid electric vehicles (PHEVs) into the electricity system of New York ISO and PJM led to declines in both CO_2 and N_2O emissions, but the contribution to SO_2 emissions was mixed (Peterson et al. 2011). This research demonstrated that although electrification of transport may not directly contribute to reduced air pollution, electric vehicle charging can contribute to emissions reductions because it is done during times of minimum system load. A smart meter and smart-charging program can reduce costs and environmental impact by timing PEV charging. Other efforts to model Grid-to-Vehicles or Vehicles-to-Grid programs highlight the role PEVs could play in grid management and electricity storage. Some believe that using electric vehicles for energy storage could also be an important innovation, as this would help the grid to integrate more varied renewables and expand consumers' access to low-cost electricity. Whether and how this ultimately will benefit consumers remains to be seen.

Other Environmental Benefits

In addition to the environmental benefits just mentioned, the changes associated with smart grid could also allow new strategies to reduce air and water pollution. Smart grid could allow for the widespread development of market tools to manage environmental emissions. For example, the Tennessee Valley Authority has developed a real-time 8,760-hour carbon intensity product which allows its industrial customers to manage the carbon-related emissions from their electricity use. If smart grid reduces fossil fuel use, other environmental benefits will include reduced environmental impacts from fossil fuel extraction such as coal mining, oil drilling, and gas fracking. Reduced fossil fuel use would also reduce NO_x and SO_x emissions, which harm human and ecosystem health. The integration of environmental management of electric system pollutants and water use into grid management could help to link the environmental impacts of electricity production to energy use and reduce environmental emissions.

2.2.4 Empowered Citizens

For some, one of the most exciting promises of smart grid is that it could empower citizens to more actively engage in the generation and management of the electricity system at multiple levels (i.e. communities, organizations, households and individuals). Citizen empowerment could transform interactions among electricity users and utilities, allow for integration of new electricity producers, and give communities and consumers more of a stake in decisions about electricity grid development and deployment.

The smart grid promise of citizen empowerment comes in many forms. Enabling and encouraging distributed generation, including household and community-level renewable electricity generation, offers new ways for citizens to engage and control their electricity systems. Providing consumers with more information about real-time electricity use through smart meters is one way to empower people to be more involved and intentional in their energy use. As consumers obtain price data from their smart meters, they are better able to manage their own energy use. The standard industry term to describe this promise is demand-side management, also discussed in section 2.2.1 on economic promise. Providing customers access to real-time price signals offers the potential for households to harmonize their consumption patterns with availability of low-priced electricity and to consider their daily practices more holistically. This information also invites individuals and households to consider options for generating their own electricity and becoming more sophisticated energy prosumers.

Taking Control

One promise of smart grid is that people who have previously been relatively passive consumers can become actively engaged in making important decisions about how they will interact with the electricity system. This promise embraces the notion that information is power: if customers have more information they can have more control, and play an active role in aligning their priorities with management of their electricity systems. In addition to better managing their personal energy use, the smart grid prosumer could be involved in the creation of community energy systems, integrating distributed generation such as rooftop solar, combined heat and power units, and demand response and community energy storage through a series of microgrids. The goal of energy independence has become increasingly important to some individuals and communities; it is not just at the national scale that energy independence emerges as a form of empowerment. To those who strive toward energy independence, smart grid offers potential for greater energy autonomy, empowering individuals, organizations, and communities to determine their own electricity generation and use (Collier 2013). Beyond the technical changes associated with prosumers generating their own electricity, prosumers are empowered to change the rules which have governed the system for the past century. Although they cannot change the laws of physics, they can push for new rules and new business models related to what matters to them. Revising existing legal frameworks to ensure component interoperability for smart grid can encourage system innovation by codifying both the new rules and the freedom to change them (Arnold 2013). As we began writing this book in the net zero-energy passive house in Donegal, Ireland designed by Cathal Stephens (Stephens 2011), we took satisfaction in the knowledge that the same wind that was chilling the outside air was also turning the wind turbine and generating the electricity to heat our water, cook our food, and power our computers.

Although, as we noted in the section on economic promises, the word *prosumer* is typically used within the energy sector to describe these individuals who simultaneously produce and consume electricity (Grijalva and Tariq 2011), these individuals have a much broader social significance. Futurist Alvin Toffler (Toffler 1980) coined the word to describe how people would function in a world made possible by electric media, where the roles of producer and consumer would become indistinguishable. More recently it has been offered as a descriptor of participants in the new world of Web 2.0, where it is possible for anyone to generate, organize, and alter information content (Gerhardt 2008).

Although prosumers may install solar panels and wind turbines to produce electricity, and then sell the excess electricity back to the grid, buying and selling electricity only scratches the surface of their influence on system change. Energy prosumers may become invested in the electricity system in ways that transcend immediate financial costs and benefits. As prosumers, they engage in all dimensions of production, including invention and experimentation. Their lack of incumbency means they have no reason to maintain outdated structures and mechanisms that lock the system into negative reproduction cycles. For many, the "electric utility death spiral" worrying many incumbents serves as a rallying cry for community energy autonomy. Instead, prosumers can imagine a smart electricity system that responds to both their personal and professional desires (Bagozzi 2008); one that has little historical baggage to thwart productive change (Grijalva and Tariq 2011). These early adopters are eager to try new technologies that seem promising for them and their communities. They gain a sense of satisfaction from involvement in the creative process, and maintain a high level of independence that may challenge the working models of some legacy actors in the energy sector. The smart grid promise of citizen empowerment could end up being the promise with the most revolutionary potential.

2.3 Pitfalls of Smart Grid

While smart grid holds many promises, it is not without potential drawbacks. This section reviews some of the perceived pitfalls of smart grid development. People with concerns over smart grid development range from those raising generic concerns about electric system vulnerabilities and offering cautionary advice to move more slowly toward grid modernization to activist opponents who imagine smart grid as ushering in a dystopic future. Skepticism and concern about an overly optimistic smart grid vision has been expressed by many. Some opponents envision a world where a vulnerable electricity sector fails to provide basic necessities such as water, food, or transportation (Investor's Business Daily Editorials 2014). Others worry that smart grid could create different levels of power quality, undermine investments in the collective power grid, and further exacerbate disparities by making high-quality energy available only to wealthy elites. For some, a smart grid is the ultimate "Big Brother," using energy data to spy on citizens in their own homes, creating a society of neurotic anxiety. This perspective views smart grid as a future panopticon, which is a building designed to allow a watchman to discreetly observe inmates at all times (Figure 2.2; Bentham 1995), invoked by Foucault as a metaphor for modern society's pervasive and invasive structures to observe and discipline all participants (Foucault

Figure 2.2 A panopticon is a design that enables constant secretive surveillance – this represents a major potential pitfall of smart grid. This image depicts the interior of the penitentiary at Stateville, United States in the twentieth century. Source: Foccault, 1995.

1995). The panopticon perspective sees smart grid as a perfect storm of corporate/ government control that will peer into private citizens' homes and usurp individual liberty and local decision-making authority. In such extreme scenarios, smart grid weakens society and the economy, and the ensuing disruptions then cause widespread environmental damage. Citizens have no motive to participate in such a system, but are better off maintaining their distance.

2.3.1 Diminished Reliability and Security

A dominant pitfall of smart grid is the possibility that instead of making the grid more reliable and resilient, smart grid has the potential to reduce system reliability. This is usually framed as an indirect result of the increased vulnerability of integrating tightly networked computerized control systems into the electricity grid. Cybersecurity risks have emerged as an increasingly pressing geopolitical issue as militaries around the world actively engage in cyber probes, attacks, and warfare. Vulnerabilities in the electricity system are the focus of increased political attention (Campbell 2011). Improved communication between electricity providers and meters at individual homes and businesses opens the door to hackers who could gain control of electric power at the household, neighborhood, or even regional level (Robertson 2009).

Compared to the traditional grid, smart grid could result in an increase in strategic vulnerabilities to a diverse array of attacks, ranging from simple jamming devices to sophisticated attacks on nuclear power plants (Levitt 2011). The vulnerability of the U.S. electricity system has been highlighted by repeated cyber probes from several countries, including Russia, China, and Iran (U.S. House of Representatives 2013). It is possible that smart grid could make electricity systems more vulnerable to the most commonly encountered risks from malware, or software used to disrupt computer operation, gather sensitive information, and gain illicit access to computer systems. Security experts warn that, in a world where the Internet penetrates every significant sector and system, the goal is not to eliminate vulnerabilities but to keep protective measures current, given the constant emergence of new vulnerabilities (Axelrod 2006).

Some of these cybersecurity risks are similar to those facing all commercial industries. For example, in 2009 hackers robbed 179,000 Toronto Hydro customers' names, addresses, and billing information from their e-billing accounts; this poses a similar challenge to the 40 million credit cards whose details were stolen from Target in 2013 (Riley et al. 2014). Other cyberattacks are particular to the electricity sector. Security consultants have demonstrated the ease with which hackers could install computer worms that could take over the entire grid. Applied to the electricity system, cyberattacks use potentially devastating tools and scenarios, including malware designed specifically to damage particular systems; hardware that can be used to insert malware into unsuspecting systems, exploiting vulnerabilities in archaic hardware; and attacks routed through third-party providers of electrical services. Although cybersecurity risks are only one among many categories of terrors that haunt contemporary society (Beck 1992), these risks are especially important in the energy sector because of the centrality of energy systems to all aspects of public life. Increasingly throughout the world, both social well-being and industrial competitiveness depend on a complex energy system that centers on electricity grids (European Commission 2013; U.S. Department of Energy 2013).

Another vulnerability relates to electromagnetic pulses (EMPs), or electromagnetic fallout that can be triggered by any major explosive burst; however, this vulnerability is associated with both the conventional electricity system and smart grid (Raloff 2009). Although EMPs do not harm humans or other life forms, they can destroy modern electronic systems by introducing massive voltage surges and shutting down vital infrastructure (Investor's Business Daily Editorials 2014). Suedeen Kelly, of the Federal Energy Regulatory Commission, and George Arnold, of the National Institute of Standards and Technology, have publicly recognized the serious risks posed by EMPs (Raloff 2009).

For some, the hazards of cybersecurity are enhanced by the impossibility of completely eliminating or even defining the risks associated with information technology in a way that is operationally useful. For example, the National Infrastructure Plan (Department of Homeland Security 2006) defines cybersecurity as "The prevention of damage to, unauthorized use of, or exploitation of, and, if needed, the restoration of electronic information and communications systems and the information contained therein to ensure confidentiality, integrity, and availability" (p. 103).

While some cyberattacks might focus on disrupting the electricity system, others concentrate on stealing confidential information and trade secrets. More recent attacks also include attempts to destroy data, control machinery, and control or disable energy networks. For example, the Industrial Control Systems Cyber Emergency Response

Team 2012 report recorded responses to more than eighty attacks in the energy sector (U.S. House of Representatives 2013). In a 2013 U.S. House of Representatives Report authored by the staffs of Representatives Markey and Waxman, more than a dozen electric utilities responded that they were under constant, daily, or frequent cyberattacks, ranging from phishing emails, to unfriendly probes, to malware (U.S. House of Representatives 2013). While smart grid could help to enhance energy security, the integration of more information communication technology into the grid could also undermine security by opening new avenues for outsiders to affect system operation.

2.3.2 Weakened Economic Conditions

Smart grid opponents also cite economic risks, primarily related to the distribution of costs and benefits. The suite of smart grid technologies may offer aggregate economic benefits but who pays for the system upgrades and who captures the benefits of the changes will depend on how the policies, regulations, and incentives are designed and implemented. Sharing the costs of smart grid investments across the energy sector, including both the wholesale power system and retail consumers, remains an important issue. Estimated costs of integrating smart grid technologies range from $27 billion for smart meters to $1.5 trillion for a fully modernized electric system. In the United States, cost estimates for deploying smart grid range from $338 billion to $476 billion (EPRI 2011).

Increased Costs for Energy Consumers

Although more efficient use of system assets should decrease electricity system costs, those lower costs will not necessarily result in lower electricity bills for consumers. Some studies show consumers paying *more* for electricity with smart grid, even though they can better manage electricity use (EPRI 2011). Changing the rate structure is hard, and while the current flat-rate structure does not reflect the true costs of electricity, it is a structure that consumers are used to. Dynamic pricing, which promises to align customer and system costs, could lower overall system costs, but it could raise costs for some customers. It has generated opposition from interest groups who believe their members' rates will increase under dynamic pricing and that it will expose consumers to additional market risks and volatility. Another pitfall for consumers relates to the initial investment to upgrade. Costs of smart grid upgrades paid for by the utility and recovered from the customers include the meter, as well as associated costs for the information communication technology (ICT) overlay, network and communication upgrades, and installation. For consumers to fully benefit from the information provided by smart meters, many will also need to install a home area network and replace old appliances with new ones. Even highly motivated consumers may face high transaction costs simply to get started using the new consumer applications.

An ethical concern related to economic risks associated with smart grid is that those who cannot afford to upgrade their home infrastructure or shift their energy use may not be able to benefit from smart grid. Additionally, low-income customers who are unable to pay increased electricity costs could also lose access. Consumer advocates caution that not everyone will benefit from smart meters and that time-of-use pricing could disproportionately affect vulnerable populations. Vulnerable groups, particularly the poor, sick, and elderly, may fall victim to price fluctuations. If electricity is a basic need, then what right do providers have to refuse it to anyone? This raises questions of energy poverty and challenges regarding incentives. What is enough and what about questions of equity? Ratemaking is essentially social policy. Regulators and utilities have explored different rate structures, or providing "sustenance" levels of electricity, to separate out the issues of energy poverty and energy management, but dynamic pricing remains a politically challenging issue.

Some question why utilities are focusing smart meter initiatives on residential customers, because large industrial customers may have the greatest ability and greatest incentive to shift demand and reduce system costs. Dynamic pricing experiments have shown that even if just a few customers reduce their electricity use a lot in response to price signals, those reductions benefit the entire system. Some argue that investments to change small-scale residential electricity usage might be better spent focusing on larger electricity users. Many residential customers are relatively small energy users and some may have limited ability to shift energy use.

Increased Risks to Electric Utilities

Electric utilities are also exposed to economic risks with smart grid. Utilities are required to reliably and affordably meet their customers' electricity demand at all times. Any system service or operational failures result in increased scrutiny. State regulations have been created to protect ratepayers and ensure adequate and reliable service. In traditionally regulated states, most regulators have not historically offered many incentives for utilities to invest in innovative new technologies. While there are exceptions to this – for example, California, Vermont, and New York Public Utilities Commissions (PUCs) were leaders in promoting energy efficiency and renewable power – most PUCs remain risk-averse. For example, the current regulatory environment in many states may not allow cost recovery for smart grid investments. Shorter anticipated lifetimes of some smart grid technologies, including software-based smart meters, are affecting utilities' ability to invest in upgraded equipment. Additionally, taking operational advantage of smart grid requires new investments in communication and network infrastructure, retraining utility personnel, and new interactions with customers. While some of these could be beneficial, change embodies risk.

For utilities in restructured regulatory environments, making a business case for smart grid investments depends upon the legacy system and competition for customers.

If a utility is in a wholesale and retail restructured environment, smart grid investments may help to keep customers, but higher costs could also drive them away.

The potential for smart grid to disrupt utilities' conventional business models is a threat that many are already feeling. Additional consumer-owned distributed generation and renewables on the system could shift system costs and utility benefits. Popularly termed the "utility death spiral," these shifting circumstances could undermine traditional utility business models, shift away from investments in large centralized plants, and raise the costs of capital for utilities. While the changing landscape presents both opportunities and risks, smart grid technologies are central to allowing this type of system change (Kind 2013; Lacey 2013; Pentland 2014). Chapter 6 provides more details on how high levels of wind and solar on the German grid have slashed revenues for traditional utilities such as RWE and Vattenfall.

2.3.3 Degraded Environmental Quality

Another potential pitfall is the possibility that smart grid could worsen, rather than improve, environmental quality. While the majority of smart grid perspectives in the media and within the energy sector highlight smart grid's potential for environmental improvement, including enabling greater penetration of low-carbon renewables, for some smart grid poses environmental concerns.

Electromagnetic Emissions May Harm Human Health

Some people are worried about human health and environmental damage caused by electromagnetic waves. Similar to concerns about high-voltage power lines in the 1990's (MacGregor, Slovic, and Morgan 1994), and worries over cell phones (Siegrist et al. 2005), a small but vocal community is concerned about the cumulative effect of exposure to radiation from electromagnetic fields (EMF) in the radio frequency (RF) band from the wireless technology used in many smart meters (Hess and Coley 2012). This concern also relates to cell phones, wi-fi networks, and other technologies which emit EMF, but many of the "Stop Smart Grid" efforts concentrate on EMF emissions from smart meters. More details of this concern are discussed in Chapter 5 on struggles with smart meter deployment.

Renewables May Destabilize the System, Causing Environmental Damage

Another environmental concern relates to the potential for high penetration of renewable resources to destabilize the electricity system, which could in some locations and some timeframes cause negative environmental impacts. Variable renewable resources require new operational protocols and lead to unanticipated environmental emissions as conventional generators are forced to rapidly ramp up and down to match demand. Resources such as large-scale wind and solar are changing the management of the high-voltage transmission system. While changes in operational protocols may be called for, if not properly managed, the additional ramping may reduce generator lifetimes and increase system costs. Additionally, as more renewables enter the grid, they alter the economic structure of the system. In Germany, this has meant that lower-carbon natural gas has been less economically attractive.

Some allege that distributed renewables such as rooftop solar PV could be affecting distribution networks and potentially compromising power quality. In locations with high levels of rooftop PV in distribution networks, such as Hawaii, California, and Australia, some utilities claim that the PV systems are potentially causing fluxes in power quality and possibly damaging appliances and electronics. This translates into an environmental concern due to inefficiencies, as well as disposability of damaged appliances.

Other Environmental Dangers

The electrification of transport enabled by smart grid could also have some negative environmental impacts if the electricity system remains carbon-intensive. If electric vehicles are charged with high-carbon electricity, they could result in more greenhouse gases than conventional vehicles. PEV could displace emissions from tailpipe to smokestack in more carbon-intensive areas. While smokestacks are usually located in less populated areas and could reduce exposure to pollutants such as nitrogen oxides and sulfur dioxide from car tailpipes, overall greenhouse gas emissions could increase. Also, unless large-scale PEV charging is carefully integrated into grid management, it could compromise system management and require additional generators, inadvertently degrading system performance.

2.3.4 Disempowered Citizens

A final category of smart grid pitfalls is the potential to disempower citizens by alienating people and compromising their privacy. For some, smart grid is only peripherally associated with modernization of the electricity system. Rather, they perceive smart grid as the future panopticon: a perfect storm of corporate/government control that will peer into private citizens' homes and usurp individual liberty and local decision-making authority. Jeremy Bentham's (Bentham 1995) *Panopticon* was designed to allow a watchman to observe all inmates of an institution without their being able to tell whether they are being watched. Although Bentham's Panopticon was primarily envisioned as the ideal design for a prison, Michel Foucault (Foucault 1995) invoked it as a metaphor for modern society's pervasive inclination to observe and discipline all participants. A panopticon creates a social environment of neurotic anxiety that knows no limits. Watchers operate in a state of heightened awareness of the perpetual riskiness of the system, while those under observation experience consciousness of their permanent visibility. In the smart grid panopticon, no walls or locks are necessary for domination.

Losing Control and Loss of Privacy

Contemporary social critics assert that smart grid enables the deployment of panoptic structures throughout society. Using smart grid technologies such as smart meters, sensors, and two-way communication, utilities or others could track users' activities, while user-generated content means that daily social activity may be recorded and shared with others, including corporate sector actors. The fear of "Big Brother" looms large in many sectors of society. Even if the panopticon model is nothing more than a marketing tool used by utilities providers in the hopes of better segmenting their markets or coming to know their customers better, it is still an invasion of privacy. Having the potential to observe everything a person does in their own home is detrimental to democratic values and raises issues of personal privacy and freedom (Jensen and Draffan 2004).

Related privacy concerns are directly associated with energy and demand management. They stem from smart meters recording and transmitting energy use data and the fear that these data will be intercepted and used by unauthorized parties to gain insights into both electricity use and individual behavior. Industrial customers may fear that electricity use information will provide competitors with information on business activities. Residential customers may believe that home energy use data provides a detailed yet unwanted window into their lives. The two major concerns are that smart meter data would reveal personal in-home behavior and that measures to protect privacy may be inadequate. How energy data are transmitted, protected and stored, and kept from misuse has far-reaching implications.

Uses of existing data provide justifications for these concerns. Monthly metered data have recently been used by law enforcement to identify suspected grow-ops or marijuana-growing operations, for example (Narciso 2011). Higher interval data could provide law enforcement with significantly more information on building occupant behaviors. This tension between law enforcement and privacy concerns links to the Fourth Amendment of the U.S. Constitution. Consumers may be worried that utilities could use the data collected through smart meters without their permission. These issues remain extremely contentious parts of some smart meter rollout programs (discussed in more depth in Chapter 5). Moreover, third-party use of data by criminals, insurance companies, or marketers could target consumers based on energy use patterns. There are concerns that even if smart meter data were anonymized, it might still identify users and track their behaviors (Scott 2009).

2.4 Conclusions

The many different promises and pitfalls of smart grid are impacting the pace and type of social and technical change. The multiple smart grid promises are attractive and exciting; they resonate with many different people for different reasons and offer hope of a better future. The multiple smart grid pitfalls represent the potential risks and tensions of social and technical change that are threatening to people with a broad range of different concerns.

The tensions between the smart grid promises and pitfalls impact incumbents and new entrants in different ways in different places, and there is great diversity in how these tensions play out among and within different organizations and groups. The promises and pitfalls of smart grid are also difficult to navigate and understand given the jurisdictional, temporal, spatial, and social complexities of electricity systems. Building on this introduction, the subsequent chapters explore in more depth these promises and pitfalls, in a variety of ways. In Chapters 3 and 4 we identify and explain smart grid technologies and the stakeholders/institutions involved in smart grid development. In Chapters 5, 6, and 7, we explore in more depth cases of specific key aspects of smart grid development, highlighting the tensions between promises and pitfalls. Chapter 8 expands to explore interactions between smart grid and climate change, highlighting specific promises and pitfalls related to smart grid's potential to help society respond to climate change.

References

Amin, M. (2013a) The Economic and Employment Opportunities of Upgrading the Power Grid. *The Institute: The IEEE News Source*. www.theinstitute.ieee.org/ieee-roundup/opinions/ieee-roundup/modernizing-the-grid-part-ii

Amin, M. (2013b) Is the Smart Grid Secure, Safe, and Private. *The Institute: The IEEE News Source*. theinstitute.ieee.org/ieee-roundup/opinions/ieee-roundup/is-the-smart-grid-secure-safe-and-private

Amin, M. and B. F. Wollenberg. (2005) Toward a Smart Grid: Power Delivery for the 21st Century. *IEEE Power & Energy Magazine*, 3, 34–41.

Arnold, G. (2013) Interoperability, Security and Privacy. Presented at *Grid School 2013*, March 7, 2013, Charleston, SC: NIST Smart Grid Program, 1–38

ASCE. (2013) American Society of Civil Engineers 2013 Report Card for America's Infrastructure. www.infrastructurereportcard.org/a/#p/home

Axelrod, C. W. (2006) Cybersecurity and the Critical Infrastructure: Looking Beyond the Perimeter. *Information Systems Control Journal*. www.isaca.org/Journal/Past-Issues/2006/Volume-3/Pages/Cybersecurity-and-the-Critical-Infrastructure-Looking-Beyond-the-Perimeter1.aspx

Bagozzi, C. X. R. P. (2008) Trying to Prosume: Towards a Theory of Prosumers as Co-Creators of Value. *Journal of the Academy of Marketing Science*, 36, 109–122.

Barzilay, M. (2013) *A Simple Definition of Cybersecurity*. ISACA. www.isaca.org/Knowledge-Center/Blog/Lists/Posts/Post.aspx?ID=296

Beck, U. (1992) *Risk Society: Towards a New Modernity*. New Delhi: Sage.

Bentham, J. (1995) *The Panopticon Writings*. London: Verso.

Berger, I. (2008) Perfecting the Power Grid. *The Institute: The IEEE News Source*, 1–3. www.theinstitute.ieee.org/technology-focus/technology-topic/perfecting-the-power-grid254

Campbell, R. J. (2011) The Smart Grid and Cybersecurity. *Regulatory and Policy Issues*, Washington, DC: Congressional Research Service. fas.org/sgp/crs/misc/R41886.pdf

Charles River Associates. (2005) *Primer on Demand-Side Management: With an emphasis on price-responsive programs*. Oakland, CA: Charles River Associates. siteresources. worldbank.org/INTENERGY/Resources/PrimeronDemand-SideManagement.pdf

Collier, S. (2013) Irreversible Trends Spur Consumer Independence: Part 1. www. theenergycollective.com/stevencollier/256461/irreversible-trends-spur-consumer-energy-independence-part-1

Congressional Research Service. (2007) H.R. 6 (110th): Energy Independence and Security Act of 2007. Washington, DC: U.S. Government Printing Office.

Danish Government. (2013) Independent from Fossil Fuels by 2050. *Denmark: The Official Website of Denmark*. denmark.dk/en/green-living/strategies-and-policies/independent-from-fossil-fuels-by-2050/

Department of Homeland Security. (2006) National Infrastructure Protection Plan. www.naruc.org/Publications/NIPP_Plan4.pdf

EPA, U. S. (2010) *Global Greenhouse Gas Emissions*. Washington, DC: US Environmental Protection Agency. www.epa.gov/climatechange/ghgemissions/global.html#two

EPA, U. S. (2013a) *Sulfur Dioxide: Health*. Washington, DC: US Environmental Protection Agency. www.epa.gov/oaqps001/sulfurdioxide/health.html

EPA, U. S. (2013b) *Overview of Greenhouse Gases*. Washington, DC: US Environmental Protection Agency. www.epa.gov/climatechange/ghgemissions/gases.html

EPA, U. S. (2013c) *Inventory of U.S. Greenhouse Gas Emissions and Sinks: 1990–2011. 505.* Washington, DC: U.S. Environmental Protection Agency.

EPRI. (2008) *The Green Grid: Energy Savings and Carbon Emissions Reductions Enabled by a Smart Grid*. Palo Alto, CA: EPRI.

EPRI. (2011) *Estimating the Costs and Benefits of the Smart Grid: A Preliminary Estimate of the Investment Requirements and the Resultant Benefits of a Fully Functioning Smart Grid*. Palo Alto, CA: EPRI.

European Commission. (2013) Research and Innovation: Energy. ec.europa.eu/research/index.cfm?pg=events&eventcode=E6E6BE44-97CD-FB0F-93D7E19E66965DAA

Foucault, M. (1995) *Discipline and Punish: The Birth of the Prison*. New York, NY: Vintage Books.

Gerhardt, W. (2008) *Prosumers: A New Growth Opportunity*. San Jose, CA: CISCO Internet Business Solutions Group.

Grijalva, S. and M. U. Tariq. (2011) Prosumer-based Smart Grid Architecture Enables a Flat, Sustainable Electricity Industry. *Innovative Smart Grid Technologies*, 1–6. tinyurl.com/orl52wr

Hertzog, C. (2013) Smart Grid: American Energy Independence. www.theenergycollective.com/christine-hertzog/246916/american-energy-independence

Hess, D. J. and J. S. Coley. (2012) Wireless Smart Meters and Public Acceptance: The Environment, Limited Choices, and Precautionary Politics. *Public Understanding of Science*, 23(6), 688–702. pus.sagepub.com/content/early/2012/11/05/0963662512464936.full.pdf

Hoag, H. (2011) Low-carbon Electricity for 2030. *Nature Climate Change*, 1, 233–235.

IEA. (2012) *Emissions from Fossil Fuel Combustion*. Paris: International Energy Agency.

Investor's Business Daily Editorials. (2014) EMP Attack On Power Grid Could Kill 9-In-10. *IBD Editorials*. news.investors.com/ibd-editorials/050914-700375-electric-grid-vulnerable-to-emp-attack.htm?ntt=power%20grid%20could

IPCC. (2007) Mitigation of Climate Change, Working Group III Report. Intergovernmental Panel on Climate Change. Cambridge, UK: Cambridge University Press.

ISO-NE. (2013) ISO Express. isoexpress.iso-ne.com/guest-hub;jsessionid=A04F2995D777EE40F662DDE96E6F06B0

Jasanoff, S. (2006) Technology as a Site and Object of Politics. In *Oxford Handbook of Contextual Political Analysis*, ed. R. G. Charles Tilly. Oxford, UK: Oxford University Press, 745–763.

Jensen, D. and G. Draffan. (2004) *Welcome to the Machine: Science, Surveillance, and the Culture of Control*. White River, VT: Chelsea Green Publishing.

Jessoe, K. and D. Rapson. (2013) *Knowledge is (Less) Power: Experimental Evidence from Residential Energy Use*. Energy Institute at Haas (EI @ Haas) Working Paper Series, Berkeley, CA.

Jones, K. B. and Zoppo, D. (2014) *A Smarter Greener Grid, Forging Environmental Progress through Smart Energy Policies and Technologies*. Santa Barbara, CA: Praeger.

Kelly-Detwiler, P. (2013) Electricity Theft: A Bigger Issue Than You Think. *Forbes*. www.forbes.com/sites/peterdetwiler/2013/04/23/electricity-theft-a-bigger-issue-than-you-think/

Kind, P. (2013) *Disruptive Challenges*. Washington D.C.: Edison Electric Insitute. www.eei.org/issuesandpolicy/finance/Documents/disruptivechallenges.pdf

Kowalenko, K. (2010) The Smart Grid: A Primer. *The Institute: The IEEE News Source*. www.theinstitute.ieee.org/technology-focus/technology-topic/the-smart-grid-a-primer545

Kurada, N., A. A. Dhanjal, and B. Venkatesh. (2013) Evolving Perimeter Information Security Models in Smart Grids and Utilities. *ISACA Journal Online*. tinyurl.com/lgz8vzq

Lacey, S. (2013) *This is What the Utility Death Spiral Looks Like*. Green Tech Media. www.greentechmedia.com/articles/read/this-is-what-the-utility-death-spiral-looks-like

LaMonica, M. (2012) Microgrids Keep Power Flowing Through Sandy Outages. *MIT Technology Review*. www.technologyreview.com/view/507106/microgrids-keep-power-flowing-through-sandy-outages/

Levitt, B. B. G. and C. Glendenning. (2011) Dumb and Dangerous: The Problems with Smart Grids. *CounterPunch*. www.counterpunch.org/2011/03/18/the-problems-with-smart-grids/

MacGregor, D. G., P. Slovic, and M. G. Morgan. (1994) Perception of Risks from Electromagnetic Fields: A Psychometric Evaluation of a Risk Communication Approach. *Risk Analysis*, 14, 815–828.

Mallett, A., R. Reiber, D. Rosenbloom, X. D. Philion, and M. Jegen. (2014) When Push Comes to Shove: Canadian Smart Grids Experiences through the Media. Paper presented at the *2014 Canadian Political Science Association Annual Conference*. Brock University, May 27–29, 2014.

Matthews, S. (2013) The Energy Fix: When will the U.S. Reach Energy Independence? *Popular Science*. www.popsci.com/science/article/2013–05/energy-gap

Narciso, D. (2011, February 28) Police Seek Utility Data for Homes of Marijuana-growing Suspects. *The Columbus Dispatch*. www.dispatch.com/content/stories/local/2011/02/28/police-suspecting-home-pot-growing-get-power-use-data.html

Nemet, G. F. and D. M. Kammen. (2007) US Energy Research and Development: Declining Investment, Increasing Need, and the Feasibility of Expansion. *Energy Policy*, 35, 746–755.

Nordman, B. (2010) Nanogrids: Evolving our Electricity Systems from the Bottom Up. www.nordman.lbl.gov/docs/nano.pdf

NRDC. (2012) The Promise of the Smart Grid: Goals, Policies, and Measurement Must Support Sustainability Benefits. NRDC Issue Brief OCTOBER 2012 IB:12-08-A.

OECD. (2012) *EIA CO2 Emissions from Fuel Combustion Highlights*. Paris: OECD.

World Energy Outlook. (2012) World Energy Outlook Factsheet. www.worldenergyoutlook.org/media/weowebsite/2012/factsheets.pdf

Pentland, W. (2014) Why The Utility 'Death Spiral' Is Dead Wrong. *Forbes*. www.forbes.com/sites/williampentland/2014/04/06/why-the-utility-death-spiral-is-dead-wrong/

Perera, D. (2012) DoD Aims for Self-reliance with SPIDERS Microgrid. *Fiercegovernment*. www.fiercegovernmentit.com/story/dod-aims-self-reliance-spiders-microgrid/2012-07-24

Peterson, S. B., J. F. Whitacre, and J. Apt (2011) Net Air Emissions from Electric Vehicles: The Effect of Carbon Price and Charging Strategies. *Environmental Science & Technology*, 45, 1792–1797.

PJM. (2013) *Demand Response Reference Materials*. www.pjm.com/markets-and-operations/demand-response/dr-reference-materials.aspx

Raloff, J. (2009) Electric Grid Still Very Vulnerable to Electromagnetic Weaponry. *Science News*, 131. www.sciencenews.org/blog/science-public/electric-grid-still-very-vulnerable-electromagnetic-weaponry.

Riley, M., B. Elgin, D. Lawrence, and C. Matlack. (2014) Missed Alarms and 40 Million Stolen Credit Card Numbers: How Target Blew It. *BloombergBusinessweek*. www.businessweek.com/articles/2014-03-13/target-missed-alarms-in-epic-hack-of-credit-card-data

Robertson, J. (2009) Security Experts Offer Caution on Smart Grids. *NBCNews.com*. www.nbcnews.com/id/32238717/ns/technology_and_science-security/t/security-experts-offer-caution-smart-grids/#.UtA3Hk2A3IU

Sandia. (2012) SPIDERS. *Energy, Climate, & Infrastructures Security*. Sandia National Laboratories. energy.sandia.gov/wp/wp-content/gallery/uploads/SPIDERS_Fact_Sheet_2012-1431P.pdf

Scott, M. (2009) How Italy Beat the World to a Smarter Grid. *Business Week*. www.businessweek.com/globalbiz/content/nov2009/gb20091116_319929.htm

Siegrist, M., T. C. Earle, H. Gutscher, and C. Keller. (2005) Perception of Mobile Phone and Base Station Risks. *Risk Analysis*, 25, 1253–1264.

Stephens, C. (2011) Tradition, Character and Energy Efficiency: An Example of Sustainable Renovation/Expansion of a Traditional Irish Cottage. Paper presented at 27th *Passive and Low Energy Architecture Conference*, Louvain-la-Neuve, Belgium, July 13–15, plea-arch.org/ARCHIVE/2011/2011_Proceedings_Vol2.pdf

TVA. (2014) *Pollution Prevention and Reduction: Carbon Dioxide*. Knoxville, TN: Tennessee Valley Authority. www.tva.com/environment/air/co2.htm

Toffler, A. (1980) *The Third Wave: The Classic Study of Tomorrow*. New York: Bantam.

Tran, M., D. Banister, J. D. K. Bishop, and M. D. McCulloch. (2012) Realizing the electric-vehicle revolution. *Nature Climate Change*, 2, 328–333.

U.S. Department of Energy. (2008) *Energy Independence and Security Act of 2007. Summary of Public Law 110–140*. U.S. Department of Energy, Washington DC. www.afdc.energy.gov/laws/eisa.

U.S. Department of Energy. (2013) Technology Development. energy.gov/oe/technology-development.

U.S. Government Printing Office. (2008) *Public Law 110-140-Dec. 19, 2007; Energy Independence and Security Act of 2007. 311*. Washington, DC: U.S. Government Printing Office.

U.S. House of Representatives. (2013) Electric Grid Vulnerability: Industry Responses Reveal Security Gaps. Washington, DC: U.S. House of Representatives.

3

Technologies of Smart Grid

3.1 Multiple Technologies and Configurations

The term "smart grid" refers to more than a single technology or even a well-defined set of individual technologies. It is an umbrella term under which multiple different electricity system technologies, both hardware and software, are developing. As we mentioned in Chapter 1, for some smart grid is characterized primarily as the addition of an information communication technology (ICT) overlay to existing infrastructure. For others, smart grid represents the installation of new transmission lines, meters, and renewable generation. The type and degree of technological change represented by smart grid varies among different societal actors. Some people view smart grid as an inevitable, already occurring evolutionary upgrade to improve existing infrastructure and reinforce the existing system. Others view smart grid as a future revolutionary shift in how electricity is generated, distributed, and used, and a potentially destabilizing change that could shift power away from incumbent actors. Recognizing the breadth of potential for both technical and social change, this chapter focuses on reviewing smart grid technologies. This technology-focused chapter connects basic engineering details with the promises and pitfalls outlined in Chapter 2. The chapter also provides technological background helpful for understanding the details of the subsequent chapters.

To represent the complexity of smart grid development, we first offer a brief description of existing electricity systems (we have labeled these "legacy systems"). We then highlight smart grid technologies and the parts of the legacy systems in which integration of these technologies has the potential to contribute to achieving the smart grid promises introduced in Chapter 2.

We begin by describing the dominant current centralized, fossil fuel-based electricity systems – that is, the legacy systems from which smart grid systems are evolving. Familiarity with the basic structure and function of today's electricity systems is critical to understand how emerging smart grid technologies might change the electricity system's function and structure. In the first half of this chapter, we present the conventional technologies for electricity supply which follows a one-way, linear path from electricity generation, to transmission, to distribution and use. In the second half of the chapter, we introduce and explain prominent emerging smart grid technologies that could enable new types of interactions and change legacy systems. Given the diversity and pace of technological change in smart grid development for the electricity sector, this review is far from comprehensive and does not include all possible smart grid technologies, but we do provide a general overview of the types of smart grid technologies that are emerging. Throughout this chapter we also explain how specific technological components connect with both the major promises of smart grid – including enhanced reliability and security, economic gain, improved environmental quality, and empowered citizen engagement – and its potential pitfalls, including decreased security, reduced privacy, and increased costs of electricity.

3.2 Our Legacy Electricity Systems

While many of us do not know much about how our electricity systems work, we do know that we need to regularly plug in our cell phones, computers, and other appliances, and we need to pay our electricity bill. Often only during power outages are we reminded of the largely invisible, complex, interconnected electricity system on which our lives have become so reliant.

Here we refer to the currently dominant, conventional electricity infrastructure as the "legacy" electric system. Our legacy electric systems are generally understood as a sequence of centralized, unidirectional steps involving four basic elements:

(1) Generation – electric energy is generated in large-scale power plants;
(2) Transmission – high-voltage electricity is transported from the power plant to substations closer to electricity consumers;
(3) Distribution – low-voltage electricity is distributed from substations to households and commercial buildings;
(4) Use – electricity is used by consumer devices like refrigerators, computers, lights, and pumps and other residential, commercial, and industrial end-use devices in homes, offices, and industries.

3.2.1 Generation in Legacy Systems

Energy is never actually "generated"; it is simply converted from one form to another. Generation of electric energy involves the conversion of chemical, mechanical, thermal, nuclear, or radiant (such as solar) energy into electric energy. The most common way to generate electric energy involves converting the chemical energy stored in fossil fuels into mechanical energy by turning a turbine to produce electric energy. In our legacy energy systems, the dominant mechanism relies on heat produced either from burning fossil fuels, splitting atoms in nuclear power, or hydropower directly driving turbines. Except for solar cells, almost all other forms of electricity generation, including fossil fuel burning, nuclear, biomass, hydro, wind, concentrated solar, and cogeneration, rely on driving a turbine to produce electricity.

Fossil fuel-based electricity generation involves harnessing the heat (thermal energy) released when the stored chemical energy in coal, oil, or gas is released during burning. This heat boils water or heats gas to create high-pressure steam that causes rotation of a wire loop through a magnetic field while it is connected to a circuit. The rotation causes a current to flow through the wire loop and through the circuit, generating electric current through a combination of rotational force and magnetics in a generator/alternator. Nuclear power involves a similar mechanism of harnessing the heat, but in a nuclear power plant the heat is produced from the splitting of atoms (fission). This heat turns a turbine to generate electricity (similar to the gas/steam-fired plants previously mentioned). Renewable-based electricity generation relies on the movement of either wind or water to turn turbines and create a current and produce electricity, the harnessing of geothermal heat or solar heat to turn a turbine, or the conversion of solar radiation, through photovoltaic cells, into electricity.

Currently, most electricity generation occurs in large, centralized fossil, nuclear, or hydro facilities. In the United States in 2013, roughly 86 percent of electricity generation came from large, centralized fossil fuel (67 percent) or nuclear (~19 percent) power plants, while renewables made up about 13 percent of total generation (EIA 2014). Regional and local variation in this electricity generation mix is high; the Pacific Northwest region has comparatively low fossil fuel reliance, with hydropower being the largest single source of generation (around 45 percent of the state of Oregon's electricity comes from hydropower), while the Southeast region of the United States is more than 95 percent reliant on fossil fuels and nuclear for electricity generation. Although wind power currently only makes up 3.5 percent of U.S. electricity generation, some states with many wind turbines generate much more than 3 percent: for example, Iowa is currently generating about 25 percent of the electricity it consumes from wind power and Texas generates more than 12 percent of its electricity from wind. The U.S. electricity mix is also dynamic, and major temporal shifts in electricity generation have occurred (Figure 3.1).

Electricity is the ultimate "real-time product" and due to limited storage options, it must be produced to meet demand. Electricity storage is currently limited and expensive, so to ensure reliable electricity service, generating capacity needs to be sufficiently flexible to meet intense and infrequent peak demand. The electricity generation capacity of a power plant/unit is the maximum level of electricity supply possible. Capacity should at all times meet base load (the amount of power required to meet minimum electricity demands) and must be sufficient to meet peak load (the amount of power required to meet maximum electricity demand) when it occurs. While base load is generally predictable and stable, the system is managed to meet fluctuations in peak demand, which varies by time of day, day of the week, month of the year, and through larger economic cycles. For example, electricity demand spikes on hot summer

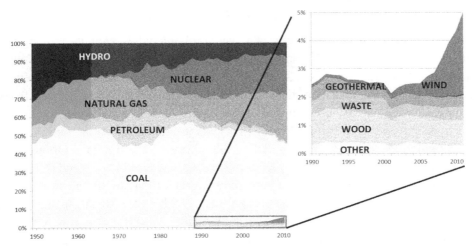

Figure 3.1 Electricity generation in the United States from 1950 to 2011 showing the percent of total electricity generated from different sources including coal, petroleum, natural gas, nuclear, and hydroelectric. The insert provides a closer view of the smaller-scale electricity generation sources. Data used in figure from EIA 2013

afternoons when people turn on air conditioning units to cool their homes. Traditionally, large coal and nuclear power plants with high capital costs but low operating costs are operated at a constant level, with minimal shutdown (only for maintenance), to meet base load. These power plants have long ramp-up times (it takes time to turn them on/off and to adjust generation), so they are unable to be particularly responsive in meeting peak demand. Natural gas plants, which are less expensive to build but were, before the shale gas boom, more expensive to run, are easier to ramp up and down and are more often used to meet shoulder load and peak demand. The need to satisfy peak-demand results in the building and maintenance of more electricity generation capacity than would be necessary if demand was more level and stable. As previously noted, industry experts compare this to sizing a church parking lot for Easter Sunday, or building a mall parking lot to cope with Black Friday customers (the day after Thanksgiving, which is the busiest shopping day of the year in the United States). Recent low natural gas prices have disrupted this order somewhat, as natural gas plants in some locations were for a short while cheaper to operate than coal facilities.

Once electricity is generated, it flows in either direct current or alternating current. Direct current (DC), electricity flowing consistently in the same direction between positive and negative terminals, flows from batteries, solar cells, and fuel cells. The electricity that is generated in a conventional power plant, on the other hand, is alternating current (AC), which means that the direction of the current reverses, or alternates, at regular intervals. The standard AC current in the United States grid is 120 reversals or sixty cycles per second (or Hertz, Hz) and 110–20 volts, while Europe's is 50 Hz and 220–40 volts. The main advantage of AC power is that voltage can easily be changed using a transformer.

3.2.2 Transmission in Legacy Systems

Electricity generation usually produces power at relatively low voltages ranging from 2 to 30 kiloVolts (kV) depending on the size of the unit, but once electricity is generated, its voltage is stepped up before transmission. A critical step between electricity generation and long-distance transmission involves a step-up transformer to increase voltage. The efficiency of transmitting electricity is greatest when the voltage is high and the current is low; higher voltage and lower current minimizes line losses (which are directly proportional to the square of the current). Often, power generation occurs far from the places where the electricity is needed, making long-distance high-voltage transmission lines a crucial part of the electric system. Long-distance transmission voltages range from 115 kV to 1200kV, so the step-up transformer plays a crucial role in increasing the voltage for transmission. Almost all long-distance transmission lines are mounted aboveground, overhead. This is primarily due to the lower cost of installing ($1–2 million per mile) and maintaining aboveground lines, which are roughly ten times less expensive than underground lines. In some urban or otherwise sensitive areas, however, transmission lines are sited belowground.

Transmission lines have typically used high-voltage three-phase alternating current (AC), although high-voltage direct current (HVDC) is increasingly being used to enhance efficiency for long-distance transmission and to reduce the potential for disruption of synchronicity due to sudden new loads or blackouts. Despite its efficiency benefits, the higher cost of HVDC transmission conversion equipment has limited the use of HVDC.

In North America there are four main interconnected regional networks of long-distance transmission lines. These interconnections provide systemwide resilience through redundancy and multiple pathways for electricity flow: they are the Western Interconnection, the Eastern Interconnection, the smaller Texas system (managed by the Electric Reliability Council of Texas), and the Quebec Interconnection (Figure 3.2). To prevent disruptions resulting from either demand exceeding supply, technology/equipment failures, or weather, regional transmission networks allow for redundancy and variation in the paths through which electricity flows.

3.2.3 Distribution and Use in Legacy Systems

High-voltage transmission lines bring electricity from generation facilities to local substations, where the power is "stepped down" to a lower voltage and then sent over distribution networks to local electricity users, including industrial, commercial, and residential customers. These substations involve a step-down transformer which reduces the voltage for distribution (generally down to 3–25kV).

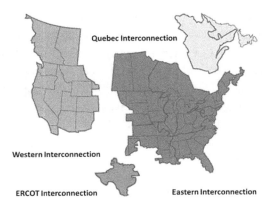

Figure 3.2 Map of the four long-distance transmission interconnections in North America. The Eastern Interconnection includes the United States and Canada, facilitating international coordination of electricity management. Source: ERCOT 2005

From the substation, electricity is distributed locally within a community to individual buildings and homes. The voltage is generally further reduced at the point of use to the standard voltage of that region, which varies in different countries (with most customers getting 110–20 V in the United States and 220–40 V in Europe) and with electricity use requirements. Electricity losses from transmission and distribution are estimated to be about 6–7 percent of electricity generated (EIA 2013), so the potential to increase efficiency by decreasing losses in transmission and distribution is high.

The use of electricity varies among different kinds of electricity consumers: industrial, commercial, or residential (Chapter 4 details different types of consumers). Demand for electricity varies over time, over the course of a day, seasonally, and annually. It is driven by consumer behavior, weather, and larger economic trends. In a typical 24-hour period, electricity demand peaks in the afternoon and early evening hours and is at its lowest in the middle of the night. Seasonal variation in electricity use is dependent on variability in the type of heating and cooling requirements of different building in different places. In many buildings, air conditioning during the summer makes this the most energy-intensive time of year. In the United States, Canada, and most EU countries, overall annual electricity demand has not been rapidly growing but has been relatively stable, although how energy is used is shifting. In general consumers are using more energy in appliances, electronics, and lighting than they did twenty years ago, powering more devices, including computers, electric toothbrushes, and smartphones.

3.2.4 Challenges Within Legacy Systems

One of the major limitations resulting in inefficiencies in legacy systems is the challenge of storing electricity. It is currently very expensive to store electricity;

large, expensive batteries limit the feasibility of storing significant amounts of electricity. Given this limited capacity for electricity storage, electricity systems need to be managed and controlled to constantly strive for real-time balancing of electricity generation and electricity demand. To ensure reliability, electricity systems assume a certain level of redundancy, and they are managed with a goal of supply capacity exceeding demand at all times. These capacity margins range from 10 to 20 percent. Electricity generation must match electricity use. If and when demand exceeds supply, rolling brownouts or disruptive power outages can occur. Power outages can trigger many different kinds of local disruptions, but they can also sometimes lead to systemwide destabilization that can result in larger regional blackouts. The Northeast United States blackout of 2003 is an example of systemwide destabilization triggered by a localized area of demand exceeding supply. When operators did not redistribute power as needed after an alarm system in an Ohio control room failed to notify them of a localized disruption, cascading systemwide impacts ensued, causing power losses for 55 million people in the United States and Canada. The interconnectivity and interdependence of electricity systems means that minor oversights in one location can result in a much larger problem.

Many electric utilities still rely on customer telephone calls to alert them to power failures. Most current electricity systems lack sensors and other technologies throughout the system that could allow system operators (as well as regulators, municipalities, or consumers) to understand how much electricity is flowing in different places in the system, or whether and where there may be disruptions. Improved sensors could alert a utility if a power failure occurred and enable them to better manage system recovery.

Another limitation of most current electricity systems relates to electricity pricing and the incentives that it creates; in most places in the United States, customers pay a flat charge per kilowatt hour (kWh) of electricity used, and the same price is charged whether the customer uses the electricity during a peak-demand time of day or in the middle of the night during a low demand time. This flat pricing fails to provide electricity users with incentives to reduce their electricity use during peak demand

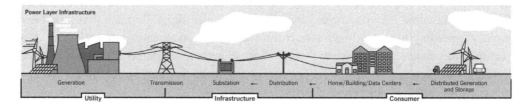

Figure 3.3 The legacy electricity system is designed to connect large-scale generation plants to load centers and is generally conceptualized in a linear model. Smart grid, distributed generation, and storage technology provide new challenges and opportunities for the system and challenge the conventional linear model.
Source: Greentech Media 2013

times. Many electricity users are unaware of the variations in demand throughout the day and are also unaware of the higher costs of generating electricity during peak demand times. Technologies informing customers about electricity use, along with a change in the price structure based on the time of day (time-of-day pricing), could help to align customer behavior to facilitate shifts in electricity use practices that could reduce the peak-load demand. Reductions in peak demand have large potential for lowering overall electricity system costs because if the maximum generating capacity is reduced, fewer power plants need to be maintained and kept online.

3.3 Technologies for Fulfilling Smart Grid Promises

Smart grid offers multiple promises to improve the legacy system and improve electricity use across society. In Chapter 2 these promises were outlined; they include enhanced reliability and security, economic benefits and cost savings, environmental improvement, and a more engaged citizenry. How can electricity system technologies contribute to realization of these promises? What technologies could induce system changes? To what degree might smart grid technologies represent a rearrangement of the current sociotechnical system?

The "smart grid" umbrella term represents the integration of digital technologies, sensors, and other ICT to empower more efficient and reliable electricity management and use. Smart grid technologies include both consumer-facing technologies (those with which consumers interact) and grid-facing technologies (those in transmission and distribution that are less visible to consumers). Smart grid technologies also include both hardware and software (Table 3.1). The next section and Table 3.1 review some of these different technologies.

3.3.1 Generation Technologies for Smart Grid

In addition to allowing better management of the legacy electric system, smart grid has the potential to better integrate different kinds of generation. Enabling the integration of renewable energy – both distributed renewable generation and large-scale renewable generation – is among the most prominent and influential smart grid promises. The specific technologies associated with this important smart grid function include specific electricity generation technologies at both the large scale and the distributed small scale, as well as the technologies required to effectively interface between the grid and renewable generation.

Although different renewable technologies are often grouped together and considered as a single set of technologies (wind, solar, geothermal, etc.), it is important to recognize that different technologies are often involved in large-scale renewable power generation and distributed-small-scale renewable installations. Technologies, policies, and activities to promote large-scale renewables are in some cases very

Table 3.1. Major smart grid technologies. Adapted from information sources from the Smart Grid Information Clearinghouse (Smart Grid Information Clearinghouse 2012).

	Stage	Technology	Definition
Generation	**Generation**	Grid-tie inverter	A grid-tie inverter (GTI) is a special type of inverter used for integrating renewable energy sources (e.g. PV) with the utility grid
Transmission	**Transmission**	Synchrophasors	Synchrophasors are "synchronized phasor measurements," that is, measurements of AC sinusoidal quantities, synchronized in time, and expressed as phasors. With a fixed temporal reference frame, synchrophasor measurements may be used to determine useful information about operation of the grid
	Transmission	FACTS/HVAC/HVDC	AC or DC (HV or LV) voltage transmission from generation substation to distribution grid
	Transmission modeling/ testing	Power transmission analysis software	Power transmission analysis software is a package consisting of tools to create, configure, customize, and manage power transmission system models which are very similar to the real-world transmission systems
	Transmission in case of HVDC distribution and generation for renewables	Inverter and rectifier	Used for AC to DC and DC to AC conversion
	Transmission	Static shunt/VAR compensator	Used for static VAR compensation as lines have mutual inductance which consumes reactive power
Transmission and Distribution	**Transmission & Distribution**	Substation automation	Automation applications: Voltage control, synchronism, load and bus transfer, load curtailment, fault detection
	Transmission, distribution & substations	Relays and breakers	Relays are signaled by CTs to detect any kind of fault in the power system and trip open the breaker to disconnect the circuit and avoid equipment damage

Category	Subcategory	Name	Description
Distribution	**Distribution (Distribution management system DMS)**	Fault locator for distribution system	Fault locators are devices and software typically installed at a substation to identify fault events, identify fault types, and calculate the distance from a monitored point to the identified fault source in a distribution system
	Distribution	Advanced metering infrastructure meters	AMI involves two-way communications with "smart" meters and other energy management devices. This allows companies to respond more quickly to potential problems and to communicate real-time electricity prices
	Distribution, Information management	Advanced substation gateway	Also known as symmetric multi-processing (SMP) gateway – an advanced computing platform that serves as a single point of access to all intelligent electronic devices (IEDs) in the distribution system
	Distribution, Information management	Distribution automation	A system consisting of line equipment, communications infrastructure, and information technology that is used to gather intelligence about a distribution system
Cross-Cutting Information Management	**Information management**	Meter data management system	Automates and streamlines the complex process of collecting meter data from multiple meter data collection technologies. Evaluates the quality of that data and generates estimates where errors and gaps exist
	Information management	RTU (remote terminal unit)	The RTU functions at the remote location wherever a SCADA system needs equipment monitoring or control
	Information management	SCADA	Computer systems that monitor and control industrial, infrastructure, or facility-based processes
	Information management	Energy management system	An energy management system (EMS) is a system of computer-aided tools used by operators of electric utility grids to monitor, control, and optimize the performance of the generation and/or transmission or distribution system

Table 3.1. (cont.)

Stage	Technology	Definition
Information management, Consumer and utility end	Smart meter	A **smart meter** is usually an electrical meter that records consumption of electric energy in intervals of an hour or less and communicates that information at least daily.
Consumer Resident/ Housing	Load control receiver	Load control receivers are devices used to control loads directly or indirectly through a low-voltage circuit such as an air-conditioner thermostat or contactor
Home area network (HAN) and load control modules (LCM): Information management	Programmable communication thermostats	A **programmable communication thermostat (PCT)** is a component of a control system which senses the temperature of a system so that the system's temperature is maintained near a desired setpoint. PCT can communicate wirelessly.
Generation/ Transmission/ Distribution/ Consumption	Short circuit current limiter	**Current limiting** is the practice in electrical or electronic circuits of imposing an upper limit on the current that may be delivered to a load to avoid damage to transmission/generation/distribution equipment
Distribution/ Generation/ Transmission transformer	Advanced on-load tap-changer	The on-load tap-changer (OLTC) is used to change the tapping connection of the transformer winding to change the voltage ratio while the transformer is still in service without interrupting the load

Consumer

Cross-Cutting

52

different than those involved in small-scale distributed generation. Large-scale wind parks, solar PV arrays, or concentrated solar-thermal projects connect directly to the high-voltage transmission grid. These generators are often directly integrated into wholesale electricity markets, with power plants sold to utilities, or bid directly into the wholesale markets. Managing large-scale renewable variability and integrating renewable resources into the system presents an ongoing challenge. Improved prediction of wind and solar resources and developing new control systems to manage large-scale renewables all depend on smart grid technologies.

By contrast, rooftop solar photovoltaic (PV) and small-scale wind turbines connect directly to the low-voltage distribution grid, which is usually operated by the local utility. Many argue that the addition of distributed generation diversifies supply, reduces risks of outages, improves overall grid reliability, and reduces fossil fuel reliance and carbon emissions, but integration into a distribution network that was not designed to directly integrate production and allow bidirectional flow has proven challenging. The Pecan Street Smart Grid pilot project in Austin, Texas, for example, has encouraged more than 200 homes to install rooftop solar PV. As more distributed generation is deployed, continued development of communication protocols and control technology is needed to ensure smooth integration. Each of these technologies could be considered a smart grid technology.

3.3.3 Transmission Technologies for Smart Grid

One smart grid technology that is crucial for the integration of more distributed renewable electricity into the grid is the grid-tie inverter (GTI) that converts DC electricity into AC. Renewable generation from solar PV or wind turbines produces DC electricity, while the grid transmits and distributes AC electricity for use by households and industrial consumers of electricity. Efficient conversion of the DC power to AC is necessary to enable more renewable generation to contribute to the electricity system for both the high-voltage transmission system and the low-voltage distribution network. With the grid-interactive inverter, electricity generated from small-scale renewable technology, such as solar and wind, which is not used on-site can be sold back to the utility's distribution network and compensated either through net metering or feed-in tariffs. The grid-tie inverter includes an oscillator that synchronizes the frequency of the renewably generated electricity to that of the grid. Another function of the GTI is to disconnect from the grid if the electricity in the grid is disrupted. This safety function prevents electricity from flowing in the downed grid system while repairs are made, but it also means that households or businesses with solar PV on their rooftops cannot use the generated power until the system is back online.

One of the most important smart grid technologies associated with improving the transmission part of the electricity system is the synchrophasor or phasor measurement unit, a technology that measures conditions on transmission lines through assessment of AC sinusoidal quantities (voltage, angle, and frequency), synchronized in time. With a fixed temporal reference frame, synchrophasor measurements may be used to monitor power flows and potentially create automatic adjustments when disruptions are identified, creating the possibility for what has become known as a "self-healing" grid. This type of monitoring also provides system operators the ability to observe the grid's overall condition (known as "wide-area situational awareness"), including the capacity to balance power flows, report outages, and receive weather, demand, and performance data in almost real time.

Among important smart grid software, Power Transmission Analysis Software is a set of tools to configure customized power transmission system models very similar to real-world transmission systems to enhance transmission system management. This type of software provides a powerful tool to enable electricity system engineers to model, design, and manage transmission networks. Among the different software available, Siemens first introduced their Power System Simulator for Engineering (PSS®E) in 1976, with version 33 released in May 2011, and Cooper Industries' CYME software has been used since 1986.

Other new management systems are also considered smart grid technologies. Energy management systems (EMS) are used by utilities to monitor, control, manage, and optimize electricity generation and transmission. A common subcomponent of EMS is SCADA (supervisory control and data acquisition) – centralized control systems that monitor and control multiple sites spread out over long distances. EMS and SCADA also facilitates smarter communication and information flow between generators, utilities, and consumers. EMS technology and SCADA is considered a part of the "legacy" system for some, while others consider it a key smart grid technology. These systems often integrate automatic control actions. SCADA represents the monitoring and control components and EMS refers to the generation control and scheduling and the overall networked system. Another tool often used with EMS and SCADA is the dispatcher training simulator which trains control-center operators.

Other sets of technologies considered by some to be part of a smarter grid includes outage management systems (OMS) that are designed to identify and resolve outages and provide historical data on past outages. Most transmission and distribution planning organizations and utilities have their own OMS that identifies real-time outages (and locations of outages) and stores a massive database of historic outages. OMS is also now used to provide real-time web-based outage information to consumers through some utility companies' websites.

Technological advances that enable higher-voltage, higher-efficiency transmission of electricity from generation substation to distribution grid are also considered among smart grid technology. An example is flexible AC transmission systems (FACTS): electronic-based equipment to control AC transmission system parameters to increase power-transfer capability. A key component of FACTS is technology to stabilize the voltage, to reduce the presence of reactive power on the lines which exists when the current and voltage are not in phase. Volt-ampere reactive (var) is a unit used to measure reactive power in an AC power system, and reactive power compensation, or what is called static var compensation (SVC), is used to regulate and stabilize grid voltage. The stabilization function of SVCs is critical to prevent voltage breakdown, particularly when new kinds of electricity generation such as wind are added to the transmission system. Substation automation applications for voltage control, synchronism, load and bus transfer, load curtailment, and fault detection are also sometimes considered to be important aspects of smart grid technology.

3.3.3 Storage Technologies for Smart Grid

The capacity to store electricity so that generation does not have to always meet realtime demand for electricity has huge potential for reducing the costs of satisfying peak demand and allowing variable renewable energy to contribute more to meeting baseload demand. Because the wind does not always blow and the sun does not always shine, integrating some kinds of electricity storage into electricity systems seems to be a necessary component of transitioning to a renewable-based system. Enhancing storage capacity is particularly important when integrating more renewable sources, because peak wind does generally not coincide with peak demand. While some energy storage technologies have been around for decades, such as the widely used pumped hydro storage, other forms of electricity storage such as compressed air storage, batteries, or flywheels have only been applied in electricity systems in a few specific instances. These different storage technologies harness either potential or kinetic energy to enable use of the energy at a later time.

Currently the U.S. electricity system only has capacity to store about 2.3 percent of total electricity production capacity, which is about 24.6GW (DoE 2013). Most of this storage (~95 percent) is from pumped storage hydro, while the combination of other storage technologies, including compressed air storage (CAES), thermal energy storage, batteries, and flywheel, make up the other 5 percent. Other regions, including Europe and Japan, have larger storage capacity in their electricity systems. Different storage technologies are more appropriate for different applications, and the benefits offered by storage are varied. Among the different services that storage provides are balancing out the time between energy generation and electricity use (loads), providing adequate reserves, damping the variability of wind and solar PV, reduction in new upgrade and new capacity build, and microgrid formation; in addition, customer storage can be used to improve reliability or save money.

With high penetration of variable resources like wind and solar, electricity storage technologies could help to enhance the flexibility, efficiency, and resilience of the grid. Energy storage is particularly valuable for leveling the costs of electricity generation because the cost of producing electricity varies considerably based on the time of day and the level of electricity demand. With storage, energy can be stored during low-cost times of day and then used to generate electricity during more expensive, peak-demand times. Storage can also help maintain power quality on the grid by providing ancillary services to ensure power quality and system function. The expansion of energy storage could minimize the need to build additional power plants and additional transmission lines to meet infrequent peak demand (California PUC 2010).

Pumped hydro is one important energy storage technology that can be applied at a large scale (100s to 1000s of MWs) and can be released quickly in response to a sudden demand for more power. Pumped hydro relies on low-cost electricity to pump water from a low-level reservoir to a higher-elevation reservoir. Once the water is at the higher-elevation reservoir it can generate electricity on demand by releasing the water back to the lower-level reservoir, letting gravity power drive the turbine to generate electricity. More than 127,000 MW of pumped hydro is in operation today globally, with over 20,000 MW in the United States.

Another approach to electricity storage that uses similar logic is compressed air energy storage, which is often used with natural gas-fired turbines. Low-cost electricity can be used to compress air to a high-pressure underground media (porous rock formations, depleted gas/oil fields, or caverns). When the pressurized air is released it reduces the amount of natural gas required to generate electricity. There are just a handful of compressed air energy storage plants in operation today, in Alabama, Texas, and Germany. Others have been proposed, but the relatively high costs and specific geological requirements have made more widespread use difficult (St. John 2013).

Battery technology is evolving rapidly and has great potential for energy storage in various types of applications. Several different large-scale rechargeable batteries are currently available, including sodium sulfur, lithium ion, and flow batteries. Energy storage batteries strive to scale up the same mechanisms used by rechargeable batteries used in cars, computers, and other applications, but remain costly.

Perhaps the most traditional type of energy storage technology is the flywheel, a massive spinning disk on a metal shaft. Electricity is used to spin the disk, and braking of the rotating disk powers an electric motor to retrieve the stored energy. The size of the disk and the speed of the rotation determines the amount of energy stored; smaller flywheels are used to stabilize voltage and frequency while larger flywheels can be used to dampen load fluctuations.

Other energy storage technologies include thermal energy storage, in which energy is stored in heat either collected in molten salts or synthetic oil, or – in the case of end-use thermal – in hot or cold storage in underground aquifers, or water or ice tanks. The ultracapacitor is another technology useful for backup power during brief interruptions; this electrical device can store energy by increasing the electric charge accumulation on its metal plates and can discharge the energy by releasing the electric charge on the metal plates. A final energy storage technology is superconducting magnetic storage, which is a winding coil of superconducting wire; changing the current in the wire can add or release energy from the magnetic field. Advantages of this technology are that energy can be stored indefinitely with minimal losses, high reliability, and low maintenance because the parts are motionless and at steady state the wires release no energy. A disadvantage of superconductors is the refrigeration required.

These energy storage technologies all have potential, but are also all associated with a wide spectrum of costs and distributed benefits among actors (the next chapter describes the main categories of different actors). The different technologies have different salience in different contexts, and there is no "one-size-fits-all" approach to valuing the benefits of these technologies (California PUC 2010).

While energy storage technologies are a potential game-changer for electricity systems and these technologies have critical potential as smart grid technologies, many remain too costly for widespread deployment. These technologies are important not only because of their impact on enabling renewable integration, but also to allow a lower overall cost of electricity. The system flexibility provided by energy storage could be valuable in multiple ways.

3.3.4 Distribution Network Technologies for Smart Grid

Multiple smart grid technologies focus on enhancing the efficiency and resilience of distribution networks. Microgrids and substation automation are two broad categories of technology that offer important smart grid potential. Microgrids are generally considered as any configuration of technologies that forms a single electrical power subsystem associated with distributed energy resources (Mariam 2013). Microgrids offer self-reliance and the potential for "islanding," which means the ability to separate completely from the larger grid under certain circumstances.

Substation automation, improving networks of communication, and enhancing remote management offer multiple benefits, including reduced operational and capital expenses, assistance in regulatory compliance, and enhanced grid security (Cisco 2011). Additional technologies include the aforementioned grid-tie inverters, which can be used on both high-voltage transmission lines and low-voltage distribution networks, and enhanced sensors, which enable outage detection and power quality management on distribution networks.

Another set of important smart grid technologies is distribution automation and FDIR (fault detection, isolation, and recovery). Investment in distribution automation has been increasing steadily as utilities see large savings in enhancing distribution networks. In terms of return on investment, some utilities expect stronger returns from distribution automation upgrades than from smart meters or other smart grid technologies (Navigant 2010). Among the important smart grid technologies relevant to distribution networks is the Advanced On-load Tap-Changer (OLTC), which is used for enhanced voltage regulation.

3.3.5 Electricity Use Technologies for Smart Grid

The most prominent smart grid technology of all is the smart meter, which is a technology focused on helping consumers and utilities to manage and monitor electricity use. In this section we introduce smart meters, but Chapter 5 provides a more detailed discussion of smart meter deployment. After introducing smart meters we will describe other electricity use technologies, including demand management technologies and smart appliances.

Smart meters, also sometimes referred to within the industry as Advanced Metering Infrastructure (AMI), allow for two-way communications between households and utilities. This enhanced communication lets utilities respond more quickly to potential problems and communicate real-time electricity prices, and allows households to adjust their electricity consumption practices based on real-time usage and pricing information. Smart meters have potential to reduce peak demand because consumers can use the information from the smart meter to lower electricity usage when demand and prices are high. By enabling more customers to participate in demand-side management programs, utilities could save money – both in the short term, by reducing energy costs, and in the long term by avoiding additional capital investments.

Smart meters also allow remote meter reading and remote monitoring of electricity consumption. This makes redundant and unnecessary the job of meter readers, individuals sent around to individual households to measure and record the amount of electricity used. Like most automation, this reduction in jobs has been viewed by some as a negative and others as a valuable cost saving. The consumer engages directly with the smart grid through the smart meter and home energy management system. This utility-supplied device (discussed in more detail in Chapter 5), and who owns and has access to the data it generates, have become important issues for consumers, state regulators, and companies. Because smart meters collect electricity use data at a much more granular level than analog meters – minute to 15-minute to hourly increments – some consumers have expressed concerns about privacy, as the smart grid could allow utilities or other third parties the ability to "see" how electricity is being used inside the home or business. To manage the vast amounts of data generated by AMI, meter data management systems have been developed to automate and streamline the process of collecting meter data from multiple meter data collection technologies and evaluate the data. Recent research on public opposition to smart meters suggests that health concerns (microwave radiation from wireless meters) have dominated the opposition in the United States and Canada, while cost and privacy concerns have played a larger role in Europe and Australia (Hess 2013).

In addition to smart meters, other categories of smart grid technologies that contribute to demand management and have potential for changing electricity consumption patterns are consumer load appliances (programmable dishwashers, water heaters, refrigerators, air conditioners, etc.) and consumer interface tools (home energy portals, web sites, cell phone apps, etc.) that allow individuals and households to see their energy use and manage it. With these integrated smart technologies, households could program washing machines or other appliances to run at the lowest-cost times or they could give control to the utility to control cycling of other appliances, such as air conditioning, refrigerators or water heaters, in return for financial incentives.

The electrification of transport, through advances in electric vehicles, is another technology that smart grid could enable. Electric vehicles have the potential to change electricity consumption patterns. In conjunction with smart meters, plug-in electric vehicles can be charged during off-peak periods, and in some configurations, researchers envision that the electric vehicle battery can serve as critical energy storage.

3.3.6 Systemwide Integrating Technologies for Smart Grid

In addition to the technologies mentioned above, smart grid offers a more holistic and integrated approach to electricity system management. One systemic challenge with legacy systems is the limited mechanisms for coordination and communication among those managing the different parts of the system. In legacy systems management of transmission and management of distribution, for example, are separate activities which occur in different parts of the utility. One of the major promises of smart grid technologies is to enable better integration of these previously separated management and operational structures. Other technological advancements that relate to systemwide integration include control system software for islanding (individual households, organizations, or communities protecting themselves from the grid) and the creation of standards. Other sets of smart grid technologies include the institutional integration of Geographic Information Systems (GIS) and protection and control technologies. Enhanced weather-prediction technology is another key technological component of smart grid, as the data from this technology facilitates integration of wind and solar power into the system (mentioned in more detail in Chapter 6).

3.4 Conclusions

In this chapter we have reviewed an array of different smart grid technologies. This review highlights the critical point that smart grid is not a single technology, but is rather a broad set of technologies offering different functionality to fulfill different priorities. For the general public, the smart meter is the individual smart grid technology that is most well-known and recognized. Our research on how the media represents smart grid shows that more than half of all articles mentioning smart grid focus on smart meters, which are a key component of the bidirectional utility–consumer relationship of smart grid. But clearly smart grid involves multiple other technologies, including renewable generation, communication software, advanced sensors, energy storage, and multiple grid-integration technologies (Table 3.1). This diversity in smart grid technologies contributes to both the flexibility of the concept and the ambiguity of the term. Different societal actors involved in smart grid development have different perspectives and priorities in relation to deploying and using these technologies. In the next chapter, Chapter 4, we provide a systematic review of these different societal actors and their dominant smart grid visions.

A few additional points emerge from this chapter on smart grid technologies. First, the diversity of different technological components means that each technology has varying levels of salience in different places and among different actors. While the smart meter is the most widely recognized individual smart grid technology for electricity consumers (discussed in more detail in Chapter 5), many other less prominent and less visible smart grid technologies also have potential for important system change. Another key point relates to dynamic linkages between smart grid technologies and the legacy system. In some instances, a specific technological change could be considered a key smart grid advance, while in another context that same technology could be considered an older, legacy technology. For example, while some utilities have been using power transmission analysis software for decades, others are only now considering this valuable transmission-management technology.

A final key point is that the many different technology configurations possible under the smart grid umbrella have potential to radically change our legacy electricity systems, but the pace and extent of changes depends on multiple social dimensions that will be discussed in subsequent chapters. How these different technologies are prioritized and deployed will determine what smart grid pathways are taken. The traditional model of a one-way flow of electricity from power plants to electricity consumers is being challenged in ways that have potential to rearrange not just the technological structure of electricity, but also the social structure of electricity. Acknowledging this interdependence of technological and social change, this chapter detailing smart grid technologies is followed by a chapter detailing a key social dimension – the dominant actors involved in electricity system development.

References

California PUC. (2010) *Electric Energy Storage: An Assessment of Potential Barriers and Opportunities*. California Public Utilities Commission, Policy and Planning Division. Staff White Paper. www.cpuc.ca.gov/NR/rdonlyres/71859AF5-2D26-4262-BF52-62DE85C0E942/0/CPUCStorageWhitePaper7910.pdf

Cisco. (2011) Substation Automation for the Smart Grid. In *White Paper*. www.cisco.com/c/en/us/products/collateral/routers/2000-series-connected-grid-routers/white_paper_c11_603566.pdf

DoE. (2013) Grid Energy Storage. U.S. Department of Energy. www.energy.gov/sites/prod/files/2013/12/f5/Grid%20Energy%20Storage%20December%202013.pdf

EIA. (2013) Independent Statistics and Analysis, U.S. Energy Information Administration. www.eia.gov/tools/faqs/faq.cfm?id=427&t=3

EIA. (2014) Electricity Supply, Disposition, Prices and Emissions. U.S. Energy Information Administration. www.eia.gov/oiaf/aeo/tablebrowser/#release=AEO2014&subject=6-AEO2014&table=8-AEO2014&region=0–0&cases=full2013full-d102312a,ref2014-d102413a

ERCOT. (2005) Launching a New Era in Texas. Electric Reliability Council of Texas. www.ercot.com/news/press_releases/show/84

Greentech Media. (2013) www.greentechmedia.com/

Hess, D. J. (2013) *Smart Meters and Public Acceptance: Comparative Analysis and Design Implications*. Paper presented at the meeting of the Sustainable Consumption Research and Action Network, Clark University, June 12–14, Worcester, Massachusetts.

Mariam, L., Malabika Basu, Michael F. Conlon. (2013) A Review of Existing Microgrid Architectures. *Journal of Engineering*. dx.doi.org/10.1155/2013/937614

Navigant (2010) Smart Grid Distribution Automation Spending to Total $46 Billion Worldwide by 2015. *Navigant Research*. Boulder, CO. www.navigantresearch.com/newsroom/smart-grid-distribution-automation-spending-to-total-46-billion-worldwide-by-2015

Smart Grid Information Clearinghouse. (2012) *Smart Grid Information Clearinghouse*. Blacksburg, VA. www.sgiclearinghouse.org/

St. John, J. (2013, July 9) Texas to Host 317MW of Compressed Air Energy Storage. *Greentech Grid*. www.greentechmedia.com/articles/read/texas-calls-for-317mw-of-compressed-air-energy-storage2

4

Societal Actors and Dominant Smart Grid Visions

4.1 Smart Grid Actors, Their Priorities, and Interactions

Many societal actors are engaged across multiple venues in planning, building, operating, and otherwise engaging with smart grid. Each actor views the promises and pitfalls of smart grid from their unique perspective, shaped by different interests, priorities, and logics about how the future energy system should function, who should control it, and who should benefit from its operation. Their perspectives also incorporate larger societal values such as equity, fairness, efficiency, control, and autonomy. The perspectives of these different actors are not fixed; they vary depending on the context and evolve over time. Nobody involved in smart grid and electricity system change works in isolation; rather, individuals and organizations engage with one another in multiple ways, navigating and negotiating across many different issues and engaging with multiple different technologies.

In this chapter we focus on the societal actors intentionally engaged in the evolution of smart grid. We explore the larger social context within which they are situated and the dynamic forces shaping their interactions with smart grid and one another. We focus on understanding the actors and their dominant visions based on a generalized review of their different priorities and perspectives. We describe four categories of societal actors involved in smart grid development: (1) utilities, energy service companies, and suppliers of electricity system equipment; (2) government entities at multiple levels; (3) consumers of electricity; and (4) civil society. Variations in the priorities and perspectives of these different actors are described in a general way throughout the chapter and are summarized in Table 4.1.

We begin with a general description of each societal actor and then attempt to situate them within a larger social and cultural context, describe their primary interests, and provide examples of how they interact with one another and with smart grid (Finnemore 1996). Based on research we have conducted over the past six years, we explore the primary interests driving each of these stakeholder communities. In the subsequent chapters, we describe in more detail frequent tensions within and interactions among

Table 4.1 Priorities and Perspectives of Societal Actors involved in Smart Grid

Key Actors	Who	Priorities and Perspectives
Electricity generation companies & private sector utilities, energy service companies and suppliers of electricity system equipment	Incumbent and new entrants to the energy field span multiple sectors and interests	Electricity companies must make a reasonable rate of return to survive and are obliged to follow applicable laws and regulations. Smart grid provides new business opportunities and challenges existing regimes
Government	National, regional, state/provincial, and local, jurisdictionally complex and varied in responsibilities	Different levels of government create policies to promote smart grid and have the responsibility to uphold and enforce laws and mandates affecting SG. Utility regulators work to ensure low-cost service and reliability and to advance government policies State or provincial level energy, natural resource and environment departments interested in climate, air quality and water use Regional organizations like RTOs are involved in planning for new capacity and transmission and managing the bulk power system and electricity markets
Consumers	Industrial, commercial and residential consumers have different patterns of energy use and abilities to use SG	A base assumption is that all consumers desire access to low-cost and reliable electricity. Some groups are also actively demanding electricity with lower environmental impacts
Civil society	Consumer advocacy, environmental, and privacy focused organizations	Civil society actors engage at different levels across the electricity system intervening in multiple venues at many levels to advance a broad range of goals (environmental, consumer protection, health, etc.).

63

these groups. We recognize that our characterization here is not comprehensive, and we appreciate the significant heterogeneity within and across societal actors. These differences are shaped by regional contexts, institutions, and individual priorities. In this chapter we present generalizable caricatures of key societal actors, and then in the subsequent case studies in the following chapters we provide more nuance and detail about how these actors interact (Cotton and Devine-Wright 2012).

Following the previous chapters that reviewed the promises and pitfalls of smart grid (Chapter 2) and the technologies of smart grid (Chapter 3), this chapter completes the first part of the book, which sets the stage for the more detailed case-specific examples in the subsequent chapters. In Chapters 5, 6, and 7 we develop more specific examples of both alignment between and tensions among these different actors, their priorities, and their perspectives. In Chapter 5 we focus on how actors interact regarding smart meter deployment, in Chapter 6 we explore actors' interactions in large-scale wind development and wind integration, and in Chapter 7 we highlight actor interactions in community-based and small-scale smart grid initiatives.

4.2 Electricity Generation Companies and Private Sector Actors

Electric utilities and other companies selling hardware and software for smart grid are directly involved in the planning, building, and operation of the electricity system. They engage directly with the technical and economic aspects of smart grid innovation. Many of these actors, including the utilities and long-standing suppliers of generation, transmission, and distribution equipment, are energy system incumbents with deep ties and long-standing relationships with the existing electric system processes and institutions. Incumbents have been directly involved in the co-creation of rules and norms of the current energy system and tend to benefit from the status quo. Newer entrants, including firms in the information and communication technology (ICT) sector and start-up companies, have different priorities as they are developing and taking advantage of novel business opportunities. They include companies selling rooftop solar PV or those combining groups of consumers to provide third-party demand response services; these actors are more likely to benefit from new rules and changes to the existing order. New and incumbent actors have different priorities and therefore have different capacities to adapt to change. Different risk tolerances are shaped by individual circumstances that in turn influence interactions with other societal actors. These differences also result in different kinds of interest in the opportunities presented by smart grid innovation.

4.2.1 Utilities

Electric power utilities generate electric power, operate the high-voltage transmission system to bring power to central substations, and run the low-voltage distribution grid to bring electric power to customers. They are responsible for planning and operating the electric power system and ensuring electricity is reliable and affordable for customers. Depending on the jurisdiction, electric utilities can be private or public companies. A utility may be "vertically integrated," serve a defined service territory, and singly fulfill all of those functions, or in restructured jurisdictions each of those services (generation, transmission, and distribution) will be provided by a separate entity. In some parts of the world utilities are owned by the public sector, while in other places they are privatized. All of these factors influence a utility's orientation toward developing smart grid.

The utility ownership models in the United States are diverse and span many of those found elsewhere in the world. In the United States, there are four main types of utilities: (1) investor-owned utilities (IOUs); (2) municipally owned utilities; (3) cooperatively owned utilities; and (4) federal power agencies. These different ownership structures shape the utility's motivations, their relationship with federal and state regulators, and their relationship with their customers.

In the United States, 193 IOUs serve 99 million customers (68 percent) and sell roughly 2,000 giga-watt hours (GWh). Traditionally, IOUs owned generation, transmission, and the distribution networks which served customers. As regulated utilities, they were granted exclusive service territories and handled customer service and billing. Infrastructure investments by IOUs were regulated by the state with a guaranteed rate of return for approved projects which was usually around 10 percent. This guarantee made it possible to attract private capital for investment, though the rate of return did shift with interest rates and other economic factors. Thirty-one states remain traditionally regulated, with "vertically integrated" utilities providing electricity to customers in exclusive service territories. The remaining nineteen states have undergone varying degrees of restructuring, which involves separating ownership of generation, transmission, and distribution networks.

IOUs are private companies governed and regulated by state and federal laws and regulations. In a traditionally regulated state, any new smart grid project or investment needs to be submitted to the appropriate state Public Utilities Commissions (PUC). If an IOU wanted to deploy 50,000 smart meters across its service territory, the IOU would prepare a proposal outlining the technical, economic, and social costs and benefits; detail any regulatory obligations the project would help fulfill; and submit it to the PUC. The PUC would open a docket on the IOU's proposal and PUC staff would evaluate the technical, economic, and social aspects of the proposal. The PUC would also hold public hearings on the project where supporters and opponents would have an opportunity to voice their concerns. Finally, after weighing all of the evidence, the Public Utility Commissioners would decide to approve or deny the project. If approved, the project can go forward and the additional costs can be included in the electricity rates which the IOU charges its customers. This allows the utility to recover its costs plus the guaranteed rate of return from its ratepayers (the electricity customers in their jurisdiction). This process, from opening a docket to final approval, can take months to years. In a restructured state, there is more variation in the utility's obligation to the PUC.

IOUs also work with and support other organizations. The Edison Electric Institute lobbies Congress on behalf of IOU interests and has developed many information materials on smart grid (Edison Electric Institute 2014). The Electric Power Research Institute (EPRI) conducts research by and on the industry and has prepared influential reports on the costs and benefits of investing in smart grid (EPRI 2011). These organizations work with IOUs to set legislative and research agendas. IOUs also interact with energy system consultants on smart grid. For example, influential reports by the Brattle Group on the future of smart grid and demand response (Fox-Penner 2010) and by McKinsey on energy efficiency (Booth, Demirdoven, and Tai 2010) have been influential in shaping popular understanding of smart grid and the role of utilities.

Municipal utilities face different constraints than IOUs in developing smart grid. More than 2000 public utilities serve 21 million customers and account for 15 percent of electric sales in the United States. While the majority of public utilities are small, others, such as the Los Angeles Department of Water & Power or the Long Island Power Authority, serve more than a million customers. Municipal utilities (sometimes referred to as munis) are not-for-profit organizations and can access tax-exempt financing to fund their projects. Munis can be organized in many ways: they can operate as a city department and either report directly to the city council or operate as an independent city agency. In some cases they are city-owned corporations and in others they work as municipal utility districts. They serve their communities directly, and revenue from their electricity sales often cross-subsidizes other municipal services such as fire protection or the police department.

Their access to capital and institutional capability to capture the benefits of smart grid also varies considerably (Fischlein, Smith, and Wilson 2009). Some municipal utilities, like Austin Energy in Texas, have been at the forefront of developing smart grid systems, with their Pecan Street Project linking more than 1,000 customers with smart meters, and including some with rooftop solar PV generation and plug-in electric vehicles (discussed further in Chapter 7). However, other munis feel constrained to limit investment in smart grid to maintain the lowest possible electric rates for their community. If a municipal utility wanted to roll out a project of 50,000 smart meters, the municipal utility manager would need to make a proposal to the city manager or city council. Local citizens could state their positions during public meetings and the city manager or council members would vote to approve or deny the project. Munis are represented in Washington D.C. by the American Public Power Association, which has lobbied for "proven and cost-effective" smart grid technologies (APPA 2014).

The 873 rural electric cooperatives cover 80 percent of the United States land area, serve 19 million customers (and 42 million people) and account for 11 percent of electricity sales in forty-seven states (APPA 2013). Rural electric cooperatives were born out of the 1930's New Deal, when 90 percent of all rural homes did not have access to electricity. Rural electric cooperatives are private nonprofit entities which are governed by a board elected by their utility customers. They are divided into those which operate the low-voltage distribution networks and manage customer sales and those that generate electricity and run the high-voltage lines (these rural cooperatives are sometimes referred to as G&Ts, which represents their focus on generation and transmission). G&Ts are governed by representatives from its member distribution cooperatives. To maintain their tax-exempt status and qualify for low-rate federal loans (rates of ~4.2–4.7 percent) for infrastructure investments, co-ops must earn 85 percent of their income. They are represented in Washington D.C. by the National Rural Electric Cooperative Association (NRECA), a powerful lobbying organization. As rural electric co-ops tend to be coal-intensive, NRECA has been vocal in its opposition to climate change legislation. For example, its members sent 500,000 comments to the U.S. Environmental Protection Agency (EPA) opposing climate regulations (NRECA 2014a).

While they are vocal and politically motivated in opposition to climate change and other environmental regulations, many co-ops see benefits in smart grid. Because they serve sparsely populated areas of the country and the cost of service per customer is expensive, rural cooperatives have invested heavily in smart metering, with 31 percent of all co-op customers using two-way smart meters (compared to 23 percent nationally; NRECA 2014b). For a smart meter project to be approved, the distribution co-op manager needs to gain the board's approval of the proposal.

A final category of utilities is federal power agencies, such as the Bonneville Power Administration (BPA) and the Tennessee Valley Authority (TVA). These organizations generate and sell wholesale electricity, but they have other responsibilities too. For example, the BPA operates hydroelectric dams on the Columbia River and coordinates with other agencies to manage flood control, agricultural irrigation, and salmon migration. The TVA's mission is even broader; in addition to electricity generation, the corporation manufactures fertilizer and promotes economic development in the area. BPA has long been involved in smart grid activities, working on the "Energy Web" concept a decade before the term "smart grid" took off and hosting several smart grid demonstration projects (BPA 2014). While TVA has developed a smart grid roadmap, it has not yet made extensive investments in smart grid technologies.

In our conversations with representatives from many different types of utilities throughout the country, we found that some utility personnel displayed defensiveness when asked about smart grid and conveyed dissatisfaction with the term. Many explained that existing electric grid operations, which provide electricity to billions of people worldwide, are already pretty smart. As utilities around the world have invested in upgrading electricity generation, transmission, and distribution systems, electric power has become – in most areas – more reliable and affordable than ever before. The utility managers we spoke with often see the suite of technologies currently labeled "smart grid" as natural "next step" technologies in a continuously advancing system. Many also view the technologies under the smart grid umbrella as a continuation of advancements that were already underway before the term was adopted. Some utility representatives also demonstrated strong appreciation for the term smart grid, explaining that it has helped utilities communicate the importance of these advancements to those outside the industry. For them, the term smart grid has enabled shared visioning of the many societal benefits of future electricity systems.

Several important forces shape an electric utility's business model. First, what is the finance structure? Is the company an investor-owned utility (IOU), rural electric cooperative, or government-owned/municipal utility? Second, does the utility operate in a traditionally regulated or restructured environment? Other important factors, such as the total amount of electricity generated and distributed (often referred to as load), whether the utility is a net importer or exporter, relationships with the state Public Utilities Commission, and the legacy power system infrastructure also affect a utility's appetite for smart grid investments and their ability to take advantage of new technological capabilities. A utility can benefit from smart grid and ensure cost recovery for investments only if the state-level regulator, municipal council, or co-op board approves the project. If an IOU, muni, or rural co-op cannot make a strong and persuasive business case and get approval for smart grid investment – whether this involves installing new smart meters or other grid-facing investment – they cannot invest in new smart grid technologies. Smart grid adoption rates, therefore, vary by jurisdiction and are often influenced by a complex set of societal factors.

The attractiveness of smart grid is shaped by the legacy power system infrastructure, and must respond to both institutional and customer needs. State-level policies and regulations provide incentives (or barriers) which can further influence smart grid investments. If a utility produces excess power for sale or needs to purchase energy to serve its customers, its smart grid needs are different. If a utility's base load is provided by coal plants or large hydro, its risks from climate legislation will be different. If a utility operates in a jurisdiction with renewable electricity requirements or policies promoting energy efficiency, these will again shape the demands on electric system development. For example, a utility with steady industrial load will not benefit as greatly from the ability to shift demand compared to a utility with a high residential load and high peak demand. Whether electric load is increasing, flat, or decreasing also determines the cost or benefit of energy efficiency and demand management investment, and the subsequent value of different smart grid technologies.

Smart grid technologies also present new risks to utilities. Increasing energy efficiency, demand-side management, and distributed generation programs could help utilities to be more productive, but they could also capture market share and shift profits from utilities. This could lead to declining utility revenues, increasing costs, and lower future profitability, which could adversely affect long-term profit projections and discourage investments in the industry (Kind 2013). How smart grid affects incumbent utilities depends on the politics and regulatory processes shaping the electricity industry. For example, German investments in solar PV and wind have changed the economics of the power industry and forced incumbents to shut down many natural gas plants (additional details are given in Chapter 6). Widespread penetration of consumer-owned distributed generation and the expansion of the prosumer (individuals involved in electricity change through their own production of electricity) could undercut electricity sales and expand regulatory decision making to include new actors with divergent interests.

As mentioned in Chapter 2, smart grid represents both an opportunity and a potential threat for utilities. While utilities gain new abilities to control system operations, there are no assurances that they will benefit economically. Under the current business model, utilities are generally paid for electricity that they generate and sell. For some utilities smart grid investments could be a way to further ensure returns. By investing heavily in costly smart grid technologies and getting them included in the rate base, they could potentially earn healthy returns in an era of decreasing or stagnant electricity sales.

4.2.2 Suppliers of Electricity System Equipment

Suppliers of electricity system equipment are another set of key actors in smart grid development. These suppliers produce the hardware and software for the electric system. Electricity systems integrate equipment for generation, transmission, and distribution of electricity; some technology suppliers want to continue to sell the equipment they have been supplying for years, while others may be eager to seize new business opportunities with new technologies and new approaches to energy management.

For companies who specialize in providing electricity system equipment, building a smarter grid presents a major business opportunity. Building and maintaining electricity system infrastructure involves complex and expensive engineering work that requires multiple types of equipment and expertise. A coal-fired power plant costs over $1 billion to build, while installing a 2 MW wind turbine costs $3–4 million. High-voltage transmission lines cost roughly $1–2 million per mile. Infrastructure investment and construction are costly in terms of hardware, software, and personnel training and many private firms are poised to capture the emerging business opportunities. EPRI estimates smart grid investment in the United States over the next twenty years could range from $338 billion to $476 billion, while providing benefits ranging from $1.2 trillion to $2 trillion (EPRI 2011).

The companies which supply smart grid hardware and software are spread across generation, transmission, and distribution networks (Figure 4.1). Some of these companies are incumbents like GE and Westinghouse. Engineering firms like ABB, Alstom, GE, Siemens, Mitsubishi Heavy Industries, UTC, and others build the hardware and control systems for electric power systems. Established ICT companies

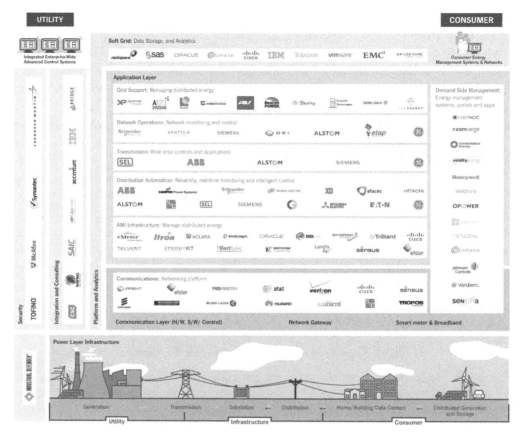

Figure 4.1 Industrial actors in smart grid, from GTM research.
Source: Greentech Media 2013

like IBM, CISCO, ORACLE, and others are also involved in smart grid equipment. For these companies, smart grid presents an enormous business opportunity, allowing them to create new products for new markets.

New entrants are also positioning themselves to take advantage of the commercial opportunities provided by smart grid development. For example, in the Northeastern United States, third-party demand response aggregators such as EnerNOC consolidate industrial and commercial customers and use the energy saved from the demand-response project to bid into electricity markets. In this way, forgone demand becomes equivalent to supply.

In Europe and the United States, companies providing detailed forecasts of wind speed, such as Energie & Meteo Systems and WindLogics, are used by wind power providers, system operations, electricity markets, and utilities to help integrate wind power into electricity systems. Other firms, such as O-Power, are working with utilities to develop consumer behavior programs and manage the consumer energy use benefits of smart grid.

Additionally, many not-for-profit industry groups like the Institute for Electrical and Electronics Engineers (IEEE) are actively involved in supporting research for smart grid development. IEEE is a professional group with 382,000 members and the organization also develops and maintains standards for electric system function, like IEEE standard 1547, which governs interconnections of distributed resources.

Different coalitions of electricity equipment suppliers are focused on different aspects of smart grid development and construction. Those involved with high-voltage transmission-line monitoring are likely involved with utilities, Federal Energy Regulatory Commission (FERC), and Regional Transmission Organizations (actors mentioned in the next section), but the lines of engagement and linkages between companies are blurry and often shifting.

Interactions between these new energy companies and incumbents have not always been smooth, highlighting tensions among smart grid actors. For example, in regions promoting rooftop solar, companies such as SolarCity are interacting with large customers such as Walmart to plan, lease, develop, and operate rooftop solar PV systems. SolarCity works with local utilities to connect their projects to the distribution network. Some utilities argue that installing solar PV generates system costs for other customers, so customers with PV systems should be subject to standby charges. Many utilities have been fighting for standby charges in PUCs and state legislatures, arguing that all customers benefit from the grid system so all should contribute to its maintenance – even if they are not purchasing electricity from the grid. SolarCity and other solar companies recently created a lobbying group, the "Alliance for Solar Choice," to fight utilities who are trying to add standby charges and change state net-metering laws which support solar installations by allowing customers to sell their excess solar power back to the grid (Herndon 2013). Solar lobbyists argue that net metering provides economic as well as public and environmental health benefits.

4.3 Government Actors

The public sector has been at the forefront of incentivizing smart grid development around the world. As seen in the earlier section on utilities, almost all electricity systems involve some level of government involvement, although the degree to which public sector actors influence, operate, and regulate electricity systems varies considerably. The following sections describe national, state, and local actors involved in smart grid development.

4.3.1 National Actors

At the national or federal level, governments support smart grid in many different ways. Government research and development programs fund cutting-edge research and demonstration projects. Government working groups convene smart grid partners to establish interoperability standards. Both federal and state legislatures pass legislation to require, incentivize, or fund smart grid projects. Federal regulatory agencies – such as the environmental and energy offices – regulate and evaluate smart grid projects. In some countries, the electric system is controlled by a national, state-owned company which finances and operates the gird. In others, federal regulations govern many aspects of the electric system, from power access, reliability, and quality standards to system costs, even if most power is provided by private companies. In some countries, such as Italy and the UK, the government has required utilities throughout the country to install smart meters.

National governments have played a particularly large role in smart grid development in the EU. The European Commission has worked with member states to develop smart grids. From 2006 to 2012, thirty EU countries have developed 281 smart grid projects, investing the equivalent of more than 1.8 billion euros. Of all EU projects, 70 percent are in just seven countries: Denmark, France, Germany, Italy, Spain, and the UK. The original and wealthier EU countries have invested more in smart grid than the newer EU entrants (JRC Scientific and Policy Reports 2013).

EU countries have different reasons for developing smart grid and are working at different paces. Italy spent 2.1 billion euros on installing 36 million smart meters from 2001 to 2008. In Italy, regulators supported smart grid development with a special tariff which provides innovative smart grid investments with an additional 2 percent rate of return for utilities. The incentive to develop a smart grid was spurred by rampant electricity theft. Sweden's investment of 1.5 billion euros to install 5.2 million meters from 2003 to 2009 was driven by a desire to create a green and sustainable energy system and develop a clean tech industry, while helping energy consumers; however, integration of data management systems remains a barrier to this last goal. Finland and Malta have also committed to full smart meter rollouts. Other large EU countries like France, Spain, and the UK have committed to full smart meter penetration by 2017, 2018, and 2019, respectively. However, some EU countries, such as Belgium, the Czech Republic, and Lithuania, are not pursuing national rollouts, and eleven other member states have not reached any official decision on smart meters and smart grid development (JRC Scientific and Policy Reports 2013).

The European Union has played an important research and coordinating role. For example, the European Commission Directorate-Generale in Research, Energy, Enterprise and Industry, Health and Consumers, Justice and groups focused on security are all actively linked to smart grid development. The Smart Grids European Technology Platform (ETP) links technology research and EU smart grid activities to national and regional smart grid initiatives, and the Smart Grids Task Force advises the European Commission on smart grid policy.

In the United States, smart grid development has linked many federal agencies and departments and created novel coalitions. Two pieces of authorizing legislation have been particularly important. In 2007, Congress passed the Energy Independence and Security Act and tasked the U.S. Department of Energy (DoE) with many smart grid coordinating activities. The legislation established the Federal Smart Grid Task Force and Smart Grid Advisory Committee and authorized the DoE to develop Smart Grid Regional Demonstration Initiatives and a Federal Matching Fund for Smart Grid Investment Costs. In 2009, Congress passed the American Recovery and Reinvestment Act, popularly known as ARRA or the stimulus bill, which provided over \$4 billion for modernizing the grid and smart grid investments. Within the U.S. federal government, the National Science and Technology Council Subcommittee on Smart Grid provides the President with recommendations on smart grid development (see Table 4.2).

Federal agencies interact with each other and the private sector. While the federal government in the United States was involved with the creation of the first interoperability standards to help ensure uniform and technology-neutral standards development, their role has evolved over time. Originally, the National Institute of Standards and Technology (NIST), DoE, and other public and private partners were tasked to create a Smart Grid Interoperability Panel (SGIP) to ensure that different smart grid components worked together and that cybersecurity concerns were addressed. The panel integrated public and private actors to work on establishing standards. In 2013, the SGIP was re-born as SGIP 2.0, and as a public–private initiative. Now funded by industry stakeholders, the re-tooled SGIP provides an open process for standards development to ensure interoperability.

Table 4.2 Important U.S. Agencies, Departments, Organizations and Programs Linked to Smart Grid.

Acronym	Name	Purpose
DoE	Department of Energy	Coordinates smart grid task force
EPA	Environmental Protection Agency	Works to ensure environmental standards are met across electric system
FERC	Federal Energy Regulatory Commission	Regulates and monitors interstate electricity transmission and markets
NERC	North American Electric Reliability Organization	Nonprofit which works to ensure electric system reliability, industry members, under FERC oversight
NIST	National Institute of Standards and Technology	Works with industry to establish technology standards
RTO	Regional Transmission Organizations	FERC-authorized voluntary organizations linking multiple states and utilities to which coordinate electricity system planning and wholesale market operations
SGIP	Smart Grid Interoperability Panel	NIST and DoE public-private program to coordinate smart grid standard development authorized under the 2007 Energy Independence and Security Act. Now SGIP 2.0
SGIP 2.0, Inc.	Smart Grid Interoperability Panel 2.0, Inc.	Nonprofit private-public partnership funded by corporations to help facilitate standards, identify testing, work on global interoperability

The DoE also interacts regularly with the private sector. It is involved in smart grid development through the direct support of research at the National Laboratories, public–private smart grid demonstration projects, and the creation of uniform standards to share energy data. For this last effort, the DoE worked with the North American Energy Standards Board, an industry consortium, to develop the "Green Button" program. This program will allow consumers to access and share energy use data with authorized third parties by establishing a common data format.

Other federal organizations are also influencing the development of smart grid through regulations (Table 4.2). The U.S. Environmental Protection Agency (EPA) regulates environmental emissions from the electricity system and sets standards for air and water emissions limits. The Department of the Interior controls federal land use and is often involved in the siting of new energy facilities or transmission projects. Concerns about cybersecurity and increased system vulnerabilities also engage the military and defense communities and bring government agencies in contact with other public and private organizations focused on system security.

4.3.2 Regional Coordination and Smart Grid

Many EU countries and U.S. states are linked together in regional electricity markets. While originally established to provide additional system reliability and improve system economics, these regional bodies have become an important force for smart grid development. These regional bodies are involved in energy markets and operations as well as long-term system planning. NordPool, which covers Norway, Sweden, Denmark, Finland, Estonia, Latvia, and Lithuania, allows for integration of renewable resources across Scandinavia. In the United States, the FERC-authorized Regional Transmission Organizations (RTOs) organize future capacity and transmission planning and market operations for roughly two thirds of bulk power sales (Figure 4.2). Chapter 6 discusses the role of RTOs in wind integration in more detail.

RTOs also play an important role in smart grid development. For example, 2009 ARRA funds provided many RTOs with funds to develop synchrophasor projects. Synchrophasors or phasor measurement units provide a fine level of data to monitor the bulk power grid and allow for enhanced system monitoring and fault detection. While RTOs cover a large portion of the United States, electricity system capacity and transmission planning and system operation continues to be dominated by states and utilities, especially in the southeast and west, with reliability enhanced by the nonprofit North American Electric Reliability Corporation (NERC) balancing authorities.

4.3.3 Subnational Actors

State and provincial governments are also actively involved in smart grid investments; their level of investment and engagement is varied. In places in the world with smaller countries, the subnational level may be important but is unlikely to have the same level

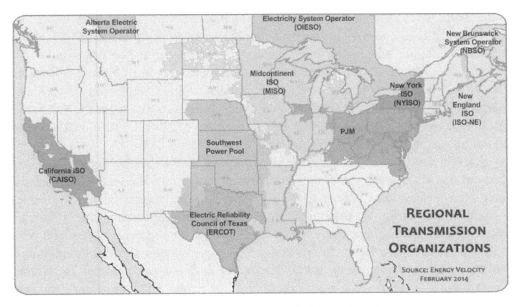

Figure 4.2 Regional transmission organizations in North America.
Source: FERC 2014

of heterogeneity as in larger countries. In the United States and Canada, states and provinces play a major role in shaping smart grid development through specific energy and environmental legislation and regulation. Smart grid development also raises important federalism issues and highlights tensions between, on the one hand, the U.S. federal government's efforts to create uniform standards and coordinate across regions, and on the other, the states, which have traditionally regulated the electric system (Eisen 2013). State legislatures pass legislation outlining broad policy directions, state PUCs approve utility plans and projects, and state energy and environmental offices often develop, evaluate, and enforce programs and regulations. State and provincial legislatures also pass renewable portfolio standards, energy efficiency legislation, and other programs to incentivize smart grid developments. State or provincial energy offices often oversee resource production, environmental compliance and energy efficiency programs. As discussed earlier, PUCs regulate IOU investments and approve utility resource plans and smart grid investments. In Ontario and Texas, the province or state has required utilities to install smart meters and enabled them to recover the costs from these investments. However, other states have not approved utility investments in smart grid and differences across jurisdictions are large.

While federal policies have been important for standard development and funding initial smart grid activities in the United States, the locus of electricity system planning, control, and regulation remains at the state level. In the U.S., several state governmental agencies play a key role in smart grid development. About half of all states have specific legislation or PUC action on smart grid development, mostly focused on meter installations (National Conference of State Legislatures 2013). In some states, such as Ohio, Vermont, West Virginia, and Washington, state legislators have passed laws encouraging smart meter and smart grid installation by requiring utilities to file plans including smart meters or by authorizing PUCs to explore and investigate smart grid technology use (National Conference of State Legislatures 2013). Public backlash against smart meter installation from groups like "Stop Smart Meters!" has led some state legislatures or county boards to pass legislation allowing customers to opt out of smart grid installation (more details on public opposition to smart meters are provided in Chapter 5).

In traditionally structured states, Public Utilities Commissioners play a crucial role for smart grid development. Commissioners and their staff of lawyers, economists, and engineers evaluate if an IOU's smart grid project provides enough consumer benefits to qualify for rate-of-return cost recovery. They also determine rate structures like dynamic pricing incentives, which affect project developers, utilities, and consumers. In addition to seeking individual project approval, in many states utilities also submit long-range resource plans to PUCs. At public hearings, Commissioners solicit input from other state agencies, nonprofits and the public, and approve or deny cost recovery for IOU smart grid projects. Smart grid innovation also requires an evolution of the traditional PUC role. Historically, the PUC mission has been to ensure adequate and reliable service and reasonable rates. Now PUCs also ensure that utilities are following state legislative mandates (for renewable power, energy efficiency, or smart grid initiatives) and other environmental regulations.

Not all state PUCs are created equally. In some cases, the PUCs are statutorily limited as to what information they are allowed to consider when approving projects or making rate-case decisions. Statutory language may not allow them to consider the benefits of multistate transmission lines on other states or to the electric system. The PUCs evaluate the legality, business case, and technical aspects of utility investments in smart grid technologies. In some jurisdictions PUCs have denied smart meter investment, finding the benefit-to-cost ratio unfavorable. This has sometimes been the case when utilities have already invested in remote meter reading and demand management with cycling of air conditioners, and cannot present a strong business case. Other PUCs have raised concerns about consumer privacy and are questioning how utilities use and manage data collected by smart meters. Additionally, the institutional capacity of PUCs varies. Commissioners may be appointed or elected, staff size and support varies, and a PUC's ability to analyze and evaluate new projects varies significantly from state to state. While the role of PUCs in the nineteen restructured states is different than in traditionally regulated states, PUCs remain key actors in smart grid deployments.

Other state-level actors include state agencies focusing on energy and natural resources, as well as those concerned with economic development, commercial interests, low-income customers, and the impact of electric rates on these groups. For example, forty-one states and the District of Columbia have a publicly supported

Rate Payer Advocate. This person works to ensure that rates remain stable and fair and represents consumer interests in rate hearings (National Association of State Utility Consumer Advocates 2013). This position was established by state legislatures in the 1970's as energy prices rapidly increased and it became clear that the setting of electricity rates often conflicts directly with other state offices pursuing other goals.

4.3.4 Local Government and Community Actors

Smart grid has the potential to empower individuals and communities to have more localized control and engagement in their energy choices. Some local governments are directly involved in electricity generation and distribution; some local-level munis and co-ops have been leaders in installing smart grid technologies. Chapter 7 focuses in more detail on local actors and the links between smart grid and community energy systems.

4.4 Consumers of Electricity

Smart grid promises consumers new ways to control and manage their electricity use, but skepticism about benefits to consumers is strong in some places. The value of smart grid to electricity users varies in part because different types of consumers are more or less able to take advantage of energy management. People consume electricity to fulfill and engage in a wide variety of societal activities. Residential comfort (heating, cooling, lighting) and function (cooking, bathing, watching TV, or gaming), commercial business use, and industrial production of goods all drive electricity consumption. Electricity is an enabling service that allows people to engage in other practices and functions, and electricity consumers need reliable, affordable energy to fulfill these needs. Some consumers are concerned about the environmental impacts of electricity generation and their electricity use, while others care more about the cost of service. How much electricity is consumed and how it is consumed varies across country, region, sector, city, and individual. Electricity use changes over time and varies across cultures, which means that smart grid has the potential to fulfill different functions for electricity consumers in different places.

While consumer interest in smart grid is often distilled by technical experts and economists into a "low-cost and reliable service," several important factors affect smart grid's salience to consumers (Cotton and Devine-Wright 2012). First, consumers use electricity differently and have different abilities to shift usage patterns. Second, consumers use electricity to power things which are embedded in complex social practices. Third, concerns about privacy or desires for more environmentally friendly power (either self-generated by prosumers or centrally supplied) also shape consumer visions of an "ideal" electricity system. These factors could change the scope of consumer control and interaction with the grid. Previously consumers could control their use by conserving energy or installing energy-efficient devices. With smart grid, consumers might have the option of actively managing their energy use profile, producing electricity from a variety of different sources, and using or selling this electricity back to the grid. Ways in which emerging smart grid technologies will interact with and shape consumer electricity use (and how changing consumer/citizen priorities will shape smart grid development) are rapidly evolving. As smart grid develops it is important to consider that energy consumption remains linked to institutional incentives, evolving cultural trends, and social practices.

4.4.1 Residential Sector: Householders and Individuals

People use energy to do things; having access to electricity is not an end in itself. Elizabeth Shove has written extensively on how attitudes, behavior and choice shape interconnected societal practices which, in turn, shape energy use (Shove et al. 2007, Shove 2010). Shove emphasizes that people engaged in social practices use things like refrigerators, electric razors, cell phones, and televisions, not energy. For example, societal attitudes toward cleanliness shape the frequency of bathing, which in turn shapes energy used to supply, heat, and dispose of water (Shove 2004). Shove also discusses how cultures shape energy use. People eat at different times across cultures and this "social synchronization of practice" shapes patterns of energy use. For example, household size, market, and shopping habits influenced by culture affect refrigerator size. Social habits shape how we build our homes, heat and cool them, and furnish them with appliances and electronics which affect electricity use. Commercial and industrial consumption also is shaped by conventions and expectations. While these practices underlie energy demand and evolve and shift over space and time, these cultural dimensions are often absent in discussions of smart grid.

Electricity use and the ability for consumers to participate in energy management is also shaped by economic factors and family and gender dynamics. Detailed smart meter data are allowing utilities to disaggregate energy use data and better understand how consumers use energy. This information allows them to create detailed household energy profiles. For example, a wealthy suburban family of six will have different energy use patterns than an elderly couple living in an urban apartment. Household energy profiles can help utilities target their communication and demand management approaches, although this profiling is a privacy concern for some consumers.

Smart grid could increase the scope of consumer engagement with the electricity system by allowing consumers to control their energy use and to produce electricity through distributed generation like rooftop solar PV or combined heat and power systems. Smart grid visionaries often talk about the emergence of the "prosumer," integrating the production of electricity with its consumption. Additionally, consumers could use electricity in novel ways, such as by driving electric cars with the electrification of transport.

In the United States, Canada, and most EU countries, overall annual electricity demand has been relatively flat in recent years, although how energy is used is shifting. Consumers use more energy in appliances, electronics, and lighting than they did twenty years ago. Although electricity is powering more devices, from electric tooth-brushes to smartphones, electricity remains affordable for most. For the average residential consumer, electricity use is often a small percentage of total household expenses. For example, in the United States residential customers spend an average of 1.3 percent of median household income in Utah to 4.6 percent in South Carolina on residential electricity (calculated from EIA data). For low-income customers, however, the proportion of income spent on energy can be much higher. Federal policymakers work with state governments, PUCs, and utilities to help low-income consumers afford energy. The Low Income Home Energy Assistance Program (LIHEAP) pro-vides federal energy assistance to families at 110–150 percent of the federal poverty level. For example, in Ohio's Percentage of Income Payment Plan Plus, consumers with incomes at or below 150 percent of the federal poverty level (roughly $35,000 in 2013) spend a maximum of 6 percent of their income on electricity (Ohio PUC 2014).

4.4.2 Commercial Electricity Consumers

Businesses use electricity differently than residential customers. For some businesses electricity is a major expense, while for others it is a minor cost. Interest in smart grid varies considerably depending on the type of business and how much electricity the company uses. Some firms may have the ability to shift energy use to different times of day and take advantage of different electricity pricing options, but others with fixed hours may not have that flexibility, making the demand management capabilities of smart grid less appealing. For example, a small independent lunchtime restaurant selling panini sandwiches toasted on an electric grill may have little ability to shift its electricity use, while a large commercial mall may have an easier time managing and reducing electricity use during specific times of peak use.

Commercial building energy use is important and some firms are strategically engaging in energy management. Ikea, Kohls, and other commercial firms are using investment in green energy to bolster their reputations. For example, Walmart has announced plans to increase renewable energy production by 600 percent from 2010 levels and plans to produce 7 billion KWh every year, with a 2020 goal of being 100 percent powered by renewable energy. The company plans to install solar PV on 1,000 rooftops (200 are under development now), as well as to decrease the energy intensity of their retail buildings by 20 percent and increase the use of energy-efficient LED lighting. The primary driver of these changes is cost savings, but the company recognizes the added bonus of the sustainability goals (Walmart 2013). Wal-mart is leasing rooftop PV systems from SolarCity and other emerging renewable energy companies.

4.4.3 Industrial Electricity Consumers

Of the different consumer classes, consumers in the industrial sector are farthest along in already reaping benefits from smart grid. Many industrial consumers already have real-time, time-of-day, or time-of-use pricing and have interruptible power contracts which allow their utilities to interrupt service in exchange for lower rates. Some industrial consumers may reduce their electricity usage by turning off equipment at certain times while others may run backup distributed generators to make up for the lost power. Some large industrial customers also generate their own electricity through distributed generation, such as combined heat and power systems or onsite renewables, and some also participate in regional electricity markets. While almost all utilities use emergency demand response with their large industrial customers, in some regions of the United States, such as those within the New England RTO or the PJM RTO, industrial customers can participate in market-based demand response programs. These programs link large industrial customers, and often third-party demand response aggregators, to RTO energy markets. The industrial customer projects its ability to reduce its electricity usage on the following day through a bid in the electricity market.

Industrial customers use energy for manufacturing and processing of goods including food, paper, chemicals, refining, iron, steel, and nonferrous metals. Non-manufacturing sectors such as agriculture, mining, and construction also rely on electricity for their industrial activity (EIA 2011). Industrial energy use varies by region, sector, and the energy intensity of the industrial processes. Electricity makes up roughly 14 percent of total industrial energy use, because many industries use liquid fuels, coal, and natural gas directly. Some energy-intensive industries which use a lot of electricity, such as aluminum smelting, have been strategically sited in areas with low electricity prices and reliable hydropower (EIA 2011). The ability of industrial customers to decrease or shift their electricity use depends on the type of process, as well as other technical and economic considerations.

Industrial customers may also worry about data privacy concerns posed by smart grid, fearing that energy use could reveal sensitive and confidential business information to competitors. While smart grid may give industrial consumers more control of their energy use, it also might present new risks.

4.4.4 Municipalities, Universities, Schools, and Hospitals (MUSH)

Another important electricity consumer group is public building managers. Public buildings and facilities have unique capabilities, limitations, and interests with regard to smart grid. Municipal and state buildings, universities, colleges, K-12 schools, and hospitals (often abbreviated as MUSH) are important electricity customers, with unique patterns of demand and use. Due to their public function, these buildings also have a demonstration quality that is different than privately owned residential or commercial buildings. Tight public sector budgets motivate these consumers to reduce electricity use and save money on energy expenditure. However, their ability to manage their energy use can be limited by their institutional capabilities, which vary significantly. For example, the University of Minnesota manages 857 buildings with 27.8 million square feet and is Xcel Energy's twelfth-largest customer. The university has an energy management office, runs its own steam generation plants, and monitors real-time electricity use in every building on the Twin Cities campus. The energy management team is actively engaged with energy management and in regular contact with the utility. Compared to a small rural hospital or elementary school, the university possesses a greater institutional capacity to take advantage of smart grid opportunities.

A final subset of actors linking consumers and utilities to manage energy is energy service companies (ESCOs). ESCOS work with all kinds of consumers to help them save energy and money. An ESCO representative will work with a company to identify potential energy savings opportunities, help arrange financing for investment, and monitor and evaluate energy savings. As many smaller companies do not have the resources or ability to identify and implement energy efficiency measures, ESCOs can create performance-based contracts and take on the financial risk of energy efficiency improvements, then share the savings with the host company. With an increased array of sensors and technologies, smart grid could potentially provide new opportunities for ESCO providers.

4.5 Civil Society Actors

Many civil society actors are engaged directly and indirectly in smart grid and energy system development. Some see smart grid as a way to further their energy or climate agendas while others view it as a potential threat to the issues they care about most deeply. Civil society actors include environmental groups, consumer advocates, and groups concerned with negative externalities of smart grid development, including privacy and negative health effects. While the opinions and orientations of civil society actors are diverse, they often engage in smart grid debates in several distinct forums. First, civic society groups work to directly shape public opinion and consensus on issues by creating position papers and public outreach/media materials, preparing educational materials, and engaging in grassroots organizing and social networking. Second, these groups also work in coalition with other actors to advocate for policies that further their goals by developing, promoting, and supporting local, state, and federal legislation. Third, civil society groups are often involved with litigation over specific projects or policies, suing utilities, government, or other actors. Finally, civil society actors can work directly with the business community in collaborative efforts to address specific issues. By working to support, shape, or thwart specific projects and developments through grassroots organizing, court action, involvement with local councils and boards, testimony at regulatory hearings, and business engagement, civic society actors are able to influence smart grid development in multiple diverse ways.

Some environmental groups are actively engaged in ensuring that smart grid development incorporates their strategic goals, including climate mitigation and reducing air and water pollution. They see smart grid as a crucial step in moving away from fossil fuel use and enabling renewable technologies to reduce greenhouse gas emissions. However, environmental groups also operate at multiple scales. National-level environmental organizations may embrace large global goals, while local chapters of the same environmental group may focus on community priorities such as local land conservation, local health concerns, or the environmental impact of specific projects. These types of differences can create conflicts within the environmental community. For example, the national organization may support the development of a large-scale solar project in the desert, but the local chapter of the same organization might oppose the installation because of the threat to endangered desert tortoise habitat.

For U.S. national-level environmental groups like the Natural Resources Defense Council (NRDC) and the Environmental Defense Fund (EDF), smart grid can help meet environmental goals. For example, EDF's smart grid fact sheet presents a broad range of smart grid promises: smart grid saves thousands of lives by reducing air pollution by up to 30 percent, allows consumers to reduce utility bills with "set-and-forget" tools to manage energy use, promotes economic growth and growth in clean energy jobs, provides a more reliable electricity service with quicker recovery times, allows for more renewable energy and less polluting fossil energy, and promotes electric vehicles (Environmental Defense Fund 2013). For the NRDC smart grid holds similar promises, which will allow it to advance its environmental and climate agenda by increasing energy efficiency and conservation, integrating more renewables, reducing CO_2 emissions, and enabling the use of electric plug-in vehicles (Succar and Cavanagh 2012). These two environmental organizations frame smart grid as a set of enabling technologies to help them meet their larger goals. They also stress the importance of embedding environmental goals within smart grid development. In addition to environmental goals, the organizations also highlight some technical and social issues for smart grid development. They discuss the need for open standards and technology neutrality to help avoid the risk of outdated technology lock-in and the need for consumer engagement, putting education as a fully funded priority, rather than a programmatic afterthought.

Smart grid has proven a more complicated issue for the Sierra Club. While the organization has come out in support of renewable energy, local chapters, which retain a great deal of autonomy, often remain divided on nearby projects and have been hostile to the construction of new high-voltage transmission lines. The Sierra Club position papers discuss the tradeoffs of smart grid and highlight the challenges of reconciling local conservation and environmental priorities with the risks of climate change and energy system transformation.

In the EU, environmental nongovernmental organizations such as Greenpeace and the European Renewable Energy Council focus on smart grid to promote large-scale renewables, energy efficiency, and demand-side management, as well as electric vehicles (European Renewable Energy Council and Greenpeace 2010). For Greenpeace this is linked to their larger energy strategy, which focuses on shutting down coal and nuclear plants and not including them in future energy systems. Greenpeace also advocates unbundling transmission grid ownership from power generation companies and ensuring data are available for independent analysis. It envisions hybrid smart grid systems which are tailored at different scales to integrate local generation in microgrids for island communities, and also support continental supergrids to enable more renewables and more international electricity trade across Europe. Greenpeace and other environmental organizations interact with member states, utilities, and other groups across Europe.

Other organizations and think tanks involved in energy and environmental policy also weigh in on smart grid. For example, the American Council for an Energy Efficient Economy, the Climate Project, and the World Resources Institute, among others, have written reports and white papers and engaged with federal, state, and local policymakers on smart grid issues. Other nonprofits, like the Regulatory Assistance Project, write about the regulatory and business model changes necessary for utilities to adopt and use smart grid technologies. Some other organizations, such as the Rocky Mountain Institute, are actively involved with technology development and experimentation in the electric sector (Rocky Mountain Institute 2011).

There are also some civil society groups directly opposed to smart grid development and smart meter installation. For example, groups like Stop Smart Meters! believe that smart grid development carries important risks, and they are engaged in fighting smart meter installation. This loose federation of groups around the United States and internationally draws members from both the right and left sides of the political spectrum. Members are concerned about privacy concerns raised by data collection and sharing, and some are also worried about the health effects of smart meters as a result of radio frequency exposure and fire hazards. Other members express worries about cybersecurity and the additional threats an interconnected electric system poses alongside its vulnerability to hackers.

Many opposed to smart meters also make the general observation that while smart grid and smart meters may help utilities (and help them increase control and profits), they may not benefit citizens. This question of who benefits from smart meters is salient to many societal actors and is discussed in more detail in Chapter 5. Stop Smart Meters! organizations are encouraging their followers to "defend their analog meters." They have protested at city halls, before county boards, and at state legislatures. This citizen involvement has led many jurisdictions, cities, counties, and states to pass legislation allowing utility customers to opt out of utility-led smart meter installations. Now, the organizations are focusing on how these opt-out programs are managed. This has become an important point of contention, as many utilities impose a service charge on customers who elect to keep their analog meters. Stop Smart Meters! groups are working to eliminate this fee and make opting out of accepting a smart meter an easier and less onerous process.

4.6 Conclusions

Each of these societal actors plays an important role in developing future smart grid pathways. Their perspectives, their ability to control the policy, decision making, and implementation processes, and their visions of evolutionary or revolutionary change in electricity systems are shaping smart grid advancement.

Among all the actors, smart grid development may offer the most direct and near-term benefits – and perhaps the greatest near-term challenges – to electric power utilities. Smart grid technologies allow an unprecedented level of system control, changing generation and grid management as well as shifting relationships with their customers. Smart grid also exposes them to new risks and vulnerabilities as more customers gain the capability to produce their own electricity. With variable renewable electricity integrated into electricity systems, utilities will be increasingly called on to ensure system reliability.

The next three chapters illustrate diverse ways that societal actors are engaging on specific aspects of smart grid development. In Chapter 5, we focus on the most publicly visible part of smart grid – the smart meter. In that chapter we explore how utilities, consumers, civil society actors, and public policymakers interact in different places to support or thwart smart meter installation. Chapter 6 then focuses on the interaction of multiple institutional actors in developing and integrating renewable wind resources into the electric grid. In Chapter 7, we explore the push to create community electricity systems and microgrids which link consumers, technology developers, and policymakers in novel ways. Each of these chapters explores the shifting roles of societal actors across smart grid development. The penultimate Chapter 8 then explores specific linkages between smart grid and climate change. Crucial questions regarding which societal actors have control and which societal actors benefit from different smart grid pathways and configurations are explored throughout the remaining sections of the book.

References

APPA. (2013) *U.S. Electric Utility Industry Statistics*. Washington, DC: American Public Power Association.

APPA. (2014) *Smart Grid*. Washington, DC: American Public Power Association.

Booth, A., N. Demirdoven, and H. Tai. (2010) The Smart Grid Opportunity for Solutions Providers. www.mckinsey.com/~/media/McKinsey/dotcom/client_service/EPNG/PDFs/McK%20on%20smart%20grids/MoSG_SolutionProviders_VF.ashx

BPA. (2014) *BPA and Smart Grid*. Portland, OR: Bonneville Power Association.

Cotton, M. and P. Devine-Wright. (2012) Making Electricity Networks "Visible": Industry Actor Representations Of "Publics" and Public Engagement in Infrastructure Planning. *Public Understanding of Science*, 21, 17–35.

Edison Electric Institute. (2014) *A Smart Grid. A Powerful Future*. Washington, DC: EEI. smartgrid.eei.org/Pages/Home.aspx

EIA. (2011) *International Energy Outlook 2011*. Washington, DC: U.S. Department of Energy.

Eisen, J. (2013) Smart Regulation and Federalism for the Smart Grid. *Harvard Environmental Law Review*, 37, 1–56. www3.law.harvard.edu/journals/elr/files/2013/05/Eisen.pdf

Environmental Defense Fund. (2013) *What Consumers Need to Know About the Smart Grid and Smart Meters*. New York: EDF. www.edf.org/sites/default/files/EDF-smart-grid-benefits-fact-sheet_0.pdf

EPRI. (2011) *Estimating the Costs and Benefits of the Smart Grid: A Preliminary Estimate of the Investment Requirements and the Resultant Benefits of a Fully Functioning Smart Grid*. Palo Alto, CA: EPRI.

European Renewable Energy Council & Greenpeace. (2010) [r]enewables 24/7: Infrastructure Needed to Save the Climate. www.greenpeace.org/seasia/ph/Global/international/planet-2/report/2010/2/renewables-24-7.pdf

FERC. (2014) www.ferc.gov/industries/electric/indus-act/rto/elec-ovr-rto-map.pdf

Finnemore, M. (1996) Norms, Culture, and World Politics: Insights from Sociology's Institutionalism. *International Organization*, 50, 325–347.

Fischlein, M., T. Smith, and E. Wilson. (2009) Carbon Emissions and Management Scenarios for Consumer-owned Utilities. *Environmental Science and Policy*, 12, 778–790.

Fox-Penner, P. (2010) *Smart Power: Climate Change, the Smart Grid, and the Future of Electric Utilities*. London: Island Press.

Greentech Media. (2013) www.greentechmedia.com

Herndon, A. (2013) Rooftop Solar Battle Pits Companies Against Utilities. *Bloomberg News*. www.bloomberg.com/news/2013-05-10/rooftop-solar-battle-pits-companies-against-utilities.html

JRC Scientific and Policy Reports. (2013) *Smart Grid Projects in Europe: Lessons Learned and Current Developments*. Brussels: European Commission.

Kind, P. (2013) *Disruptive Challenges*. Washington, DC: Edison Electric Institute.

National Association of State Utility Consumer Advocates. (2013) *The History of NASUCA*. Silver Spring, MD: NASUCA. nasuca.org/about-us/

National Conference of State Legislatures. (2013) *States Providing for Smart Metering*. Washington, DC: NCSL. www.ncsl.org/research/energy/states-providing-for-smart-metering.aspx

NRECA. (2014a) *Electric Co-ops Send 500,000 Comments to EPA Opposing Climate Regulations*. Washington, DC: NRECA. www.nreca.coop/electric-co-ops-send-500000-comments-epa-opposing-climate-regulations/

NRECA. (2014b) *Smart Grid*. Washington, DC: NRECA. www.nreca.coop/nreca-on-the-issues/energy-operations/smart-grid/

Ohio PUC. (2014) *PIPP Plus*. Columbus, OH: Ohio PUC. www.puco.ohio.gov/puco/?LinkServID=07FFEA58-039D-EDDB-0E00B39F911A7503

Rocky Mountain Institute. (2011) *Reinventing Fire Electricity Sector Methodology*. Snowmass, CO: Rocky Mountain Institute.

Shove, E. (2004) *Comfort, Cleanliness and Convenience: The Social Organization of Normality* (New Technologies/New Cultures). New York: Bloomsbury Academic.

Shove, E. (2010) Beyond the ABC: Climate Change Policy and Theories of Social Change. *Environment and Planning A*, 42, 1273–1285.

Shove, E., M. Watson, M. Hand, and J. Ingram. (2007) *The Design of Everyday Life*. Oxford: Berg.

Succar, S. and R. Cavanagh. (2012) *The Promise of the Smart Grid*. New York: NRDC. www.nrdc.org/energy/smart-grid/files/smart-grid-IB.pdf

Walmart. (2013) Walmart Announces New Commitments to Dramatically Increase Energy Efficiency and Renewables. www.news.walmart.com/news-archive/2013/04/15/walmart-announces-new-commitments-to-dramatically-increase-energy-efficiency-renewables

5

Smart Meters: Measuring, Monitoring, and Managing Electricity

5.1 Tensions in Metering

"You can't manage what you can't measure."

Philip Drucker

"Not everything that can be counted counts, and not everything that counts can be counted."

Albert Einstein

These two famous maxims, the first attributed to the management scholar Philip Drucker and the second to Albert Einstein, represent a tension associated with measurement and data. Drucker's quote is often used to justify the need for frequent assessment and monitoring of everything from worker productivity to consumer confidence, while Einstein's reminds us of the risks of overemphasis on generating and analyzing data.

Within energy systems, the Drucker phrase has been used to justify the need for devices to improve management of electricity demand through improved real-time measurement and monitoring. The value of enhanced measurement and monitoring in electricity systems has been increasingly recognized as a way to more closely link the costs of generation with consumer behavior, particularly by the utilities that manage electricity distribution. Access to consumer-level data on electricity use has the potential to more effectively meet the many new pressures facing electric utilities, including enhancing efficiency, reducing emissions, increasing reliability, and accelerating recovery from disruptions. This chapter explores the tension between the ideal that measurement could revolutionize household energy management and use and the experience that suggests a more incremental impact.

Consumer or household-level meters that monitor, record, and transmit electricity-use information to utilities are often referred to as "smart meters." Smart meters provide a critical link enabling bidirectional communication between electricity consumers and electricity system managers. The meter provides a way for utilities to communicate real-time costs of electricity to users while also allowing consumer use information to be transmitted to the utility. While smart meters are the tangible hardware that consumers see, they also rely on rapidly evolving software to enable the bidirectional communication.

In industry parlance, smart meters and their associated sensors and links to the distribution network are called "Advanced Metering Infrastructure" (AMI). Rather than using the term "smart meter," many policy and technical documents refer to AMI technology that bidirectionally transmits and receives information on energy use. AMI is different than one-way Automatic Meter Reading (AMR), which refers to technological advancements that enable automatic meter reading and reduce the need for meter readers to visit individual consumers each month. AMR includes both wired and wireless technology, with different levels of automation limiting the need to manually record the meter, although some AMR still requires periodic meter-reading visits. For example, with AMR, meter readers only have to drive by a house and the meter will be read remotely.

Meters are the most visible part of the grid system. As we explained in Chapter 3, smart grid includes multiple technologies with many different combinations and configurations. Some smart grid technologies are "grid-facing" and behind-the-scenes, not visible to the general public, but smart meters are the public face of the smart grid. For example, only grid operators interact with the synchrophasor technologies for monitoring the phase angle on high-voltage transmission lines with potential to improve system reliability. But other smart grid technologies, such as solar PV and smart meters, are more visible to electricity consumers. Smart meters have become the most publicly prominent type of smart grid technology, and they have taken on a symbolic role, in a sense representing the broader challenges and larger opportunities associated with smart grid development. Electricity meters are where the electricity system interfaces with the customer, and they are the part of smart grid that has potential to directly influence individuals' patterns and expectations of electricity use. The meter is a portal, where households and individuals interface with the rest of the grid, so the smart meter is a critical focal point. When our research team analyzed media reporting on smart grid coverage in national-level newspapers we found that smart meters were the technology mentioned in over half (58 percent) of news articles about smart grid (Langheim 2013).

This chapter tells several stories related to smart meter deployment; these stories highlight the controversies, tensions, and social complexities of sociotechnical change. Most often, smart meter deployment in the United States has been initiated and implemented by electric utilities as they attempt to integrate new monitoring and management strategies in response to different pressures, such as increasing system reliability and efficiency or allowing more distributed renewable energy resources onto the system. While utilities have successfully installed smart meters in millions of households and communities throughout the United States, Canada, Europe, and other parts of the world, in some places these installations have incited power struggles between utilities and concerned citizen groups. Public concerns related to privacy, health, safety, and costs have resulted in resistance that has stalled or halted some projects and resulted in legally mandated moratoriums in others (for example, smart meter moratoriums are in effect in Santa Cruz, California, and multiple municipalities in British Columbia). Strong public opposition has also forced some communities and state legislatures to institute opt-out policies which allow customers to choose whether or not to use a smart meter. This resistance has been expensive and frustrating for many utilities. Some utilities invested in smart meters in response to certain societal pressures, only to have their smart meter projects thwarted by other kinds of societal pressures.

In this chapter, we first explain what makes a meter "smart." Not all smart meters are created equally; variations in AMI and different consumer interfaces have developed with different levels of functionality. We then review the history, current status, and trends of smart meter deployment, characterizing the extent of households that have smart meters. Next we present the perceived risks and concerns of smart meters that citizens in some communities have raised. We then explore in more depth the power struggles involved in smart meter deployment in several different places, including California (Pacific Gas and Electric's smart meter rollout and the Sacramento Municipal Utility Department smart meter program), Massachusetts (National Grid's smart meter pilot in Worcester), and the national strategy for smart meter deployment in Germany. We end this chapter with a discussion of the policies and timescales for smart meter deployment and possible trajectories of future smart meter development.

5.2 What Makes a Meter "Smart"?

Meters are devices that measure electricity usage at each household or business. Conventional analog electricity meters monitor accumulated electricity usage mechanically with the turning of a dial. Analog meters, which have not changed much in the past hundred years, require utilities to hire meter readers, who travel from house to house to periodically read the meter and record the electricity usage documented on the meter. Once the meter is read, the utility generates an electricity bill which reflects the amount and cost of electricity used by the customer the previous month.

Smart meters include real-time sensors to measure electricity usage at sub-hourly intervals. Smart meters also monitor the power quality and immediately notify the utility if the power goes out. With this technological advance, many new electricity management strategies and technologies become possible. For one, smart meters eliminate the need for an electric company employee, a meter reader, to physically come and read the meter. Smart meters also reduce the likelihood of electricity theft, automatically inform the power company of any disruptions in power, and provide critical information to help restore power after an outage. Without smart meters, utilities continue to rely on phone calls from customers to notify them of power outages.

Beyond enabling remote and constant measurement of electricity and communication back to the utility, smart meters and their consumer displays can also provide new tools for electricity consumers to manage their electricity use or operate home area networks. The hope is that this energy use information will encourage customers to increase their energy efficiency, reduce and shift electricity demand, and lower their costs. In-home displays can provide households with detailed electricity use information and, if coupled with real-time pricing information from the utility, this could represent a key benefit of smart meters. In theory, users will have information on real-time electricity costs and be able to shift their electricity use. Thus a smart meter with dynamic pricing seeks to align incentives and mechanisms to change consumer energy use behavior. Theoretically, this will encourage consumers to shift non-time-sensitive electricity use from peak to nonpeak times. For example, smart meters could be linked with smart appliances which respond automatically to signals from the utility to enable, for example, the water heater or refrigerator to automatically reduce cycling during peak electricity periods, or the dishwasher start time could be shifted to coincide with lower-cost electricity periods during the middle of the night, when the least expensive generators are producing electricity. These links between the electric system and other smart appliances within the house, business, or industrial facility can create multiple opportunities to automate demand response and reduce peak usage.

In some smart meter projects, the installation of smart meters also includes in-home displays, and the in-home display is often assumed to be a critical part of AMI. But in other projects, in-home displays are not included and not considered a standard part of smart meter deployment; some smart meter programs rely solely on web-based and mobile applications for communication back to electricity users, and others provide no information at all (Weiss et al. 2009). These deployment differences affect consumers' ability to manage energy use. Recent smart meter research done in experimental "smart homes" in Karlsruhe, Germany suggests that in-home displays are an effective way to raise awareness about in-home energy use (Paetz, Dutschke, and Fichtner 2011), yet different smart meter initiatives have different views on the value of installing in-home displays. For example, within National Grid's Smart Energy Solutions pilot project in Worcester, Massachusetts, only a select subset of the pilot project participants will receive in-home displays. While outdoor smart meters have been installed in 15,000 residences and all of these participants will be able to log in to an online site to review their energy use, only about 3,000 of the participating households, about one fifth, will receive in-home energy management display technology.

These differences in consumer engagement and interpretations of what makes a meter "smart" have led to customer confusion and disjointed expectations. A recent

Carnegie Mellon study on expectations of smart meters found that many Americans have unrealistically optimistic perceptions about the potential benefits of smart meters because they assume that in-home displays and energy management abilities are fundamental aspects of smart meters, when actually most smart meters being installed in the United States do not include this service (Krishnamurty et al, 2014). Many smart meter installation programs are adopting a strategy of focusing on installing meters to replace the analog meters with an assumed intent that in-home displays can be added on later or purchased by the customer independently. This strategy may have backfired in some places, because when consumers do not experience any direct benefits from initial smart meter installation, resistance and opposition are more likely to emerge. In-home displays appear to be a basic, standard piece of most smart meter rollouts in Europe, while in the United States many utilities have replaced analog meters outside the home with smart meters without necessarily installing an in-home display, or even informing the residents of the system change. The Carnegie Mellon researchers suggested that utilities expand their smart meter programs to include in-home displays so that more consumers will experience direct benefits and real change in how they interact with the electricity system (Krishnamurti et al., 2012). Additional research points out that it is not just a matter of whether or not an in-home display is present; the type, style, and functionality of the in-home display has potential to influence the degree to which householders use the information provided by the smart meter and change their energy consumption habits (Weiss et al., 2009). Without the tangible benefit of information provided by in-home displays to inform their energy use, many customers do not experience or recognize the benefits of smart meters; skepticism persists and sometimes even grows, especially among those who feel the meters are being forced upon them.

Another key technological promise of smart meters is their ability to connect to and automatically control appliances within the home to enable demand management of electricity use over time. For example, the smart meter could be linked to the hot water heater, refrigerator, washing machine, and/or the clothes dryer and could remotely control these appliances to ensure they only run when demand and electricity costs are lowest (probably in the middle of the night). This potential for smart meters to connect to appliances is frequently mentioned in media coverage of smart grid; our media analysis research found that in newspaper articles about smart grid, consumer appliances and smart meters were the two smart grid technologies most often mentioned together (Langheim 2013). Although the capacity for smart meters to reduce in-home electricity use simply by automatically adjusting the timing of appliances in the home is mentioned prominently in media coverage about smart grid, actual deployment of smart meters with these consumer interface tools and sale of accompanying (and often more costly) "smart appliances" that enable this level of control are still limited across the United States (Navigant Research 2014).

When considering the environmental and economic smart grid promises associated with more efficient use of electricity, the capacity of smart meters to influence how individuals and households use electricity emerges as a critical component of smart grid. Smart meters are often considered the critical technology, with huge potential to promote energy conservation practices and change patterns of electricity demand. These high expectations of smart meters are based on conventional wisdom which often assumes that electricity consumption is determined by rational individual choice and that information and economic incentives are key to influencing those individual choices (Darby 2006). Recent research challenges these assumptions by providing new understandings about the social complexities of behavior change and electricity use (Brown 2014). We now know that electricity consumption patterns are embedded in social practices that integrate many drivers beyond rational individual choice; supplying information and creating economic incentives for making different choices will not necessarily lead to straightforward and intended behavior changes (Shove, Pantzar, and Watson 2012). Despite this awareness among those who are advancing knowledge about the relationship between technology and social change, smart meter deployment efforts continue to assume that information and data have a high potential for shifting and reducing electricity consumption patterns.

Beyond differences between Europe and the United States regarding the prevalence of in-home displays in smart meter programs, another difference relates to methods of communication – that is, whether smart meters use wireless or wired technology. Due to differences in existing IT and electric infrastructure – in particular, more undergrounding of electricity wires in Europe – more European smart meters rely on communication along hardwired power lines or power line communication (PLC), which does not require wireless technology. In the United States, however, where there is less undergrounding of power lines, smart meters most often rely on wireless communication. It is this reliance on wireless that has been the source of much of the health concerns associated with smart grid, related to the electromagnetic radiation emitted through wireless communication (Hess 2013).

The smart meter is not a singular, static technology. Different types of smart meters are being installed in different places and the technology is rapidly evolving as new designs emerge to suit different contexts. In addition to bidirectional communication, which is the basic function of a smart meter, many smart meters now include the capacity to automatically reduce load during peak times and to disconnect and reconnect to the grid remotely, and some now interface with gas and water meters. Most of these devices can connect to mobile applications so that users can monitor and manage their household electricity use on their mobile devices.

As we next consider the history, status, and trends of smart meter deployment, it is also important to consider that smart meter deployment is not necessarily a straightforward process. Complicated regulatory structures that dictate how electric utilities can recover costs from investments in new technologies result in a complex web of incentives. Historically utilities have invested in technologies that they amortize over decades. Many software-based smart meter technologies have much shorter lifetimes, so utilities are struggling with the question of how to adjust their expectations for timeframes of investment in new technologies.

5.3 History, Status, and Trends of Smart Meter Deployment

Over the past fifteen years, smart meters have been deployed throughout the world. Among the earliest leaders in smart meter installations were Enel in Italy and Ontario Hydro in Canada; these utilities achieved widespread deployment in their jurisdictions by 2005 and 2010, respectively. Many early deployments were supported by government funds. In the United States, for example, the 2009 American Recovery and Reinvestment Act (ARRA) invested government funds for energy innovation, including $4.5 billion to invest in advancements in smart grid (DoE 2010). As an economic stimulus package, these ARRA funds were required to be spent quickly to encourage near-term economic recovery and jobs, so among the many different technologies associated with smart grid, smart meters were appealing ready-to-go, off-the-shelf, immediately available technology. Some projects were criticized as being hastily rolled out. In hindsight some speculate that this time-crunched deployment may have unintentionally led to negative responses from some communities, where utilities gave insufficient time or attention to engaging residents and addressing their concerns.

In response to the controversy and public opposition that emerged from the attempts to rapidly deploy smart meters in 2009 and 2010, many recent smart meter initiatives have been deliberately proactive in community education and engagement. Sophisticated multiphase pilot projects have been developed to facilitate a process of learning-by-doing and preparing households for installation and engagement with new smart meter technology. The details of one pilot project, National Grid's Smart Energy Solution Program in Worcester, Massachusetts, will be discussed in more depth later in this chapter.

The total number of installed smart meters has been steadily increasing around the world, with North America leading the way with the highest overall penetration percentage (that is, the percentage of total electricity meters that are smart meters). By the end of 2013 it seems that at least 25 percent of all electricity meters in the United States and 22 percent in Europe were smart meters (Berg Insight 2013). There is variation in estimates for overall smart meter penetration rates per country because different estimates draw from different data sets and/or have different definitions of what constitutes a smart meter.

In the United States, a total of 46 million smart meters had been installed by July 2013, and 65 million smart meters are estimated to be installed nationwide by 2015 (FERC 2013). Some states, such as Texas and Maine, have high levels of smart meter

Table 5.1 Smart meter rollout plans in European countries.
Source: Eurelectric 2013

Country	Start Date	Current Status
Italy	2000	Complete
Sweden	2003	Complete
Finland	2009	Complete
Malta	2010	Complete
Spain	2011	In progress
Austria	2012	In progress
Poland	2012	In progress
Ireland	2012	In progress
Estonia	2013	In progress
France	2013	In progress
Romania	2013	In progress
Norway	2014	In progress
Great Britain	2014	In progress
Netherlands	2014	In progress
Ireland	2012	In progress
Denmark		Being negotiated
Luxembourg		Being negotiated

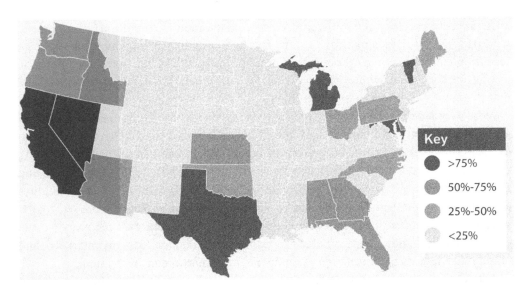

Figure 5.1. Estimates of smart meter penetration by the end of 2014.
Source: Greentech Media 2014

deployment, while others such as Montana and Louisiana have minimal penetration (Figure 5.1). The regional variation in smart meter deployment reflects the heterogeneity of the United States' energy landscape, and the reluctance of some Public Utilities Commissions (PUC) or utilities to pursue smart meter projects.

This heterogeneity is also prevalent in Europe, where there are stark differences in smart meter rollout programs among countries (Table 5.1). For example, smart meters have been installed in more than 90 percent of households in Italy, driven largely by the state-controlled power provider Enel's effort to reduce illegal and informal grid connections, or electricity theft. Italy spent 2.1 billion euros on installing 36 million smart meters from 2001 to 2008. In Italy, regulators supported smart grid development with a special tariff which provides innovative smart grid investment with an additional 2 percent rate of return for utilities. Sweden's investment of 1.5 billion euros to install 5.2 million meters from 2003 to 2009 was driven by a desire to create a green and sustainable energy system and develop a clean tech industry, while helping energy consumers; however, integration of data management systems remains a barrier to this last goal. By early 2014, Finland had exceeded its smart meter installation target: close to 100 percent of homes now have smart meters (Electric Light & Power 2014). Other countries have also committed to full smart meter rollouts: France, Spain, and the UK have committed to full smart meter penetration by 2017, 2018, and 2019 respectively. However, some EU countries, such as Belgium, the Czech Republic, and Lithuania, are not pursuing national rollouts, and eleven other member states have not reached any official decision on smart meters and smart grid development (JRC Scientific and Policy Reports 2013).

Based on the EU countries that have developed roadmaps for full deployment of smart meters, projections estimate 60 percent penetration (170 million smart meters) by 2019 (M2M Research 2013). The EU has established an ambitious goal of installing smart meters in 80 percent of households by 2020, but Germany's decision to opt out because it would be too costly for consumers has reduced the likelihood that this goal will be met (Johnston 2013).

When considering how, when, and why smart meters are being installed, there is great heterogeneity and variation. While EU installations are being promoted as part of an EU-mandated goal of 80 percent of households having smart meters by 2020, installation in the United States has been shaped by state policy, PUC decisions, and individual utility initiatives, with different levels of deployment across communities and regions. In the United States, many utilities must have projects approved by the state PUC or similar state-level regulatory authority to secure cost recovery, so the existing technologies, price of electricity, projected demand, energy markets, and many other factors determine the attractiveness of smart meter investment.

While much of the recent media coverage surrounding smart meters has focused on residential electricity use, sophisticated metering has already been used extensively in many industrial facilities and commercial businesses. Industrial electricity customers are often subject to different rates structures than residential customers, which often include a demand charge as well as charges for electricity used. Many also have interruptible power contracts which allow utilities to suspend service in emergency situations. Also, many commercial and industrial facilities manage their electricity use much more carefully than residential customers, because electricity costs can be a substantial portion of their operating expenses. Metering mechanisms for measuring and monitoring electricity use among commercial and industrial customers, therefore, may not be as novel as metering in the residential sector and may provide more direct benefits.

5.4 Risks and Concerns: Opposition and Resistance

In addition to the pressures and incentives affecting electric utilities' decisions to invest in smart meters, understanding the heterogeneity of smart meter installations requires considering public concerns associated with meter deployment. We mentioned several of these concerns in Chapter 2 where we reviewed both the overarching promises and the pitfalls of smart grid systems. Given that smart meters are the most prominent, publicly recognizable part of the electricity system, many of the perceived pitfalls of smart grid are also viewed as pitfalls of smart meters.

Before exploring the struggles and tensions of smart meter deployment in a few specific places, this section reviews the dominant health, privacy, safety, and cost concerns that smart meter opponents have raised when confronted with utility efforts to deploy smart meters. Recent research characterizing public opposition to smart meters points out that from a policy perspective there are at least two ways to respond to the resistance: (1) opposition can be viewed as a communication failure between the utilities and residents; or (2) opposition can be viewed as an opportunity for innovation in the overall system design (Hess 2013). Either way, opposition is often linked to public concern about larger societal issues.

It is important to contextualize opposition to smart meters: most new technologies encounter skepticism and some level of resistance upon initial deployment. Risk perception research highlights that we often do not perceive risks in a rational way (Slovic 2006). Instead, risk perceptions are shaped and often amplified by social factors that influence our cognitive processing (Pidgeon, Kasperson, and Slovic 2003). With new technologies, research has shown that individuals who have been involved or informed in decisions related to implementing a novel technology are more likely to minimize potential risks and view the technology favorably, while those who are not involved or informed are more likely to maximize potential risks and view the same technology unfavorably. These differences can be further exacerbated when individuals feel they have little power or influence over the direction of technological change. Instead, people may feel like a new technology – such as a smart meter installed by their utility – is being imposed on them, which may heighten sensitivity to any potentially negative aspects and fuel resistance to the technological change.

Public opposition to smart meters has been justified by multiple concerns which vary by jurisdiction. In some communities, the most prominent concern is health impacts from electromagnetic radiation, while in others the most critical concern relates to higher costs or incorrect billing or the loss of privacy given the vast amounts of household-specific electricity data being collected. Skepticism about effective management and the appropriate use and sharing of the vast quantities of data collected by smart meters has grown in response to increasing societal concern about surveillance of communications. Recent research found that Europeans are more concerned with privacy risks, while North Americans (in both the United States and Canada) are more concerned with health risks (Hess 2013). Safety concerns regarding smart meters have also emerged and are related to several electrical fires that were started following faulty smart meter installation; the apparent cause of fires reported in Pennsylvania, California, and British Columbia was either defective meters or faulty installation (Clarke 2012). Additionally, concerns about excessive costs and inaccurate billing highlight the possibility that instead of achieving the promise of saving consumers money, the meters could raise costs and provide only minimal benefits to consumers.

5.4.1 Health Concerns

Health concerns of smart meters are related primarily to the uncertain and not well-characterized risk of exposure to electromagnetic fields (EMF) radiation in the radio frequency (RF) band from the wireless technology used in many smart meters (Hess and Coley 2012). This type of radiation, often referred to as RF EMF, is non-ionizing, meaning that unlike higher-energy, higher-frequency radiation such as X-rays or uranium decay, RF EMF does not have enough energy to directly damage DNA inside living cells. While ionizing radiation is known to cause cellular disruption, the impacts of non-ionizing radiation are less certain; it is known that non-ionizing radiation does have a thermal effect in that at certain levels of exposure the radiation heats up living tissue, but whether this translates into damage remains uncertain (Rivaldo 2012). The frequency and power of the RF waves of smart meters is similar to that of cell phones and Wi-Fi routers. Compared to cell phones, RF exposure from smart meters is less because smart meters communicate short, pulsed messages throughout the day and smart meters are generally installed outside the home, so the source of the RF waves is farther away from people than a cell phone, which is held close to the head (American Cancer Society 2012). The strength of RF diminishes with distance from its source (WHO 2006), so many experts do not view exposure to RF from smart meters as a significant risk. Exposure standards and guidelines for RF exposure have been set by both national and international organizations, but significant variation exists in these standards. In the United States both government agencies (the Federal Communication Commission) and industry associations (including the American National Standards Institute and the Institute of Electrical and Electronics Engineers) have been involved in establishing standards. Partially in response to a growing movement of activists calling for a revision and tightening of existing RF standards in countries throughout the world (Behari 2012), the World Health Organization (WHO) has begun a process to harmonize standards across countries (World Health Organization 2014).

Not all smart meters emit RF; in Europe and in some Canadian jurisdictions where there is more extensive undergrounding of wires, many smart meters rely on hardwiring or fiber for the bidirectional communication. But in the United States and other places where there is minimal undergrounding of electricity wires or a lack of fiber networks, smart meters use wireless communication which is associated with RF. A more general controversy surrounding the health impacts of EMF RF has been growing in recent years as wireless technology, including cell phones, wi-fi, cordless phones, cellular antennas, and towers, have become ubiquitous. A new term, "electromagnetic smog," represents this concern about increased exposure to low-level radiation resulting from the explosion of wireless technologies. Connecting to the widespread societal awareness about the health impacts of air pollution, this phrase implies substantial health risks, demonstrated in this quotation from an report posted online by an independent researcher: "'electromagnetic smog' … like real smog … can have serious effects on our health" (Goldsworthy 2007). Some individuals now self-identify as being particularly sensitive to electromagnetic radiation, with a broad range of symptoms including headaches, nausea, dizziness, and sleep problems (Behari 2012).

Smart meter installation over the past decade has coincided with the rapid expansion of all kinds of wireless technologies, so smart meters are just one among many new devices that are associated with this relatively new perceived risk. A series of recently published synthesis reports have reviewed the science on the health impacts of EMF RF, with the majority of these reports concluding that the risks are minimal (EPRI 2010; California Council on Science & Technology 2011; Rivaldo 2012); however, some remain concerned about this emerging set of ill-characterized risks (Behari 2012). For smart meters in particular, the health risks appear lower than in the case of other technologies such as cell phones or indoor wi-fi systems, but smart meters continue to receive significant attention. While the benefits of cell phones and wi-fi are clear to consumers, individuals often feel that smart meters are new technologies forced on them by the utilities and that they do not benefit them directly. These controversies surrounding RF are similar to those surrounding high-voltage transmission lines, and in some ways similar to recent controversies about uncertain scientific claims and risks associated with vaccines, fluoride in drinking water, and even genetically modified organisms (Slovik 2010; Hess 2014). Each of these societal struggles involves redefining expert and lay perceptions of risks that are difficult to unequivocally prove or deny. These struggles also involve a strong sense among a cohort of the population that new risks are being forced upon them without their consent, and that they have limited options for refusing to adopt or use the new technology.

5.4.2 Privacy Concerns

Beyond the perceived health concerns posed by smart meters, some citizens have also expressed concern about the protection of privacy. Smart meters gather electricity use data at the sub-hourly level and when utilities have access to household-specific electricity usage data, some fear a loss of privacy. These detailed data collected by smart meters and transmitted to utilities present a new type of personal surveillance, coming on the heels of multiple privacy-related scandals and mounting pressure to reduce widespread surveillance and better protect individuals' privacy (ACLU 2014). Electric utilities will have access to information about detailed electricity use, which could reveal energy use behavioral patterns inside a home that could identify periods of vacancy and periods of intensive occupancy. While this detailed consumption data would allow utilities to better manage electricity use and better target programs and is one of the main promises of smart meters, utilities or law enforcement could also tell when electricity usage is suspiciously high or potentially being stolen (Darby 2010).

Expectations for privacy and the role of government to protect privacy vary around the world. In Germany, for example, the government is entrusted with protecting people's privacy and preventing the private sector from gaining access to personal information; a strong and coherent set of laws focus on protecting privacy and there is a privacy "czar" who is a government official (Keller 2008). In the United States, on the other hand, most citizens are more worried about the government invading their privacy than the private sector, so expectations of government protection are low and privacy laws are inconsistent and negotiated through state PUCs and legislatures. Recent revelations about the United States' National Security Administration recording and monitoring phone calls and emails and massive private sector data theft, including cases mentioned in Chapter 2, have heightened U.S. citizens' awareness of privacy concerns and data vulnerabilities posed by smart meters. Another emerging concern relates to data ownership and access. While utilities often keep energy use data confidential, many third-party providers or state officials would like to access these data for product development or energy efficiency program evaluation.

Beyond the health and privacy concerns of smart meters, some opponents also have safety concerns associated with a perceived risk from electrical fires. These concerns emerged from a few specific instances of electrical fires during or shortly after installation (EMF Safety Network 2014).

5.4.3 Cost Concerns

In addition to health, privacy, and safety, another fundamental criticism of smart meters relates to costs. The cost of a smart meter varies from about $60 to about $500, but this cost is dependent on scale (how many smart meters are being purchased) and location (different regions of the world have different markets). Also, like any modification of a component embedded within a larger system, the full cost of a smart meter program is associated with many additional factors, including the current state and maintenance needs of the preexisting infrastructure – including the analog meters – and the state of existing utility–community relationships and communication mechanisms.

Deploying these devices to all customers is expensive, and some customers may be unable to benefit from the energy management potentials provided by smart meters. While some customers may be easily able to shift their energy use, others may have a harder time. Cost concerns are coupled with equity concerns; while industrial and commercial customers may be able to capture savings through management of electricity consumption, the picture is less clear for residential customers. Consumer advocates worry that residential customers – especially low-income customers – will pay for the smart meter investment, but that the benefits will flow to electric utilities. These advocates point out that the utilities benefit from cost recovery of the deployment of large numbers of costly smart meters, but that small residential consumers may not be able to recoup the benefits to justify the costs. Partly in response to these kinds of concerns related to equity and who benefits, General Electric made a decision to manufacture smart meters in its manufacturing facility in the South Side of Chicago to create jobs and contribute to local benefits (Bomkamp 2013). There is also skepticism among some regarding the utility company's motives for installing new meters. Rather than viewing the meters as a tool for their household to manage energy use and save money, for some the meters represent another tool for the utilities to increase their profit and potentially extract more money from their customers. Some electricity users distrust the electricity suppliers' claims about the predicted costs and benefits.

Indeed, when utilities first touted the benefits of smart meters in industry-focused publications, many of the benefits focused on lower costs for the utilities – less money needed for meter readers, the ability to remotely disconnect non-paying customers, increased managerial efficiency. During these initial discussions the benefits to customers were not emphasized as much. The tone and emphasis has now shifted to focus on consumer benefits, but long-term customer savings have not yet been well quantified.

Concern about the breakdown of costs and benefits has resulted in some communities and regions retreating from initial smart meter installation goals. Germany is the most prominent example; despite the EU Energy Directive's requirement that 80 percent of European households have smart metering by 2020, the German Federal Ministry of Economic Affairs released a report in 2013 concluding that the costs of comprehensive deployment of smart meters were greater than the potential benefits (Berst 2013). The impact of this report on deployment of smart meters in Germany and beyond is still not clear, but it does highlight the struggles and controversies affecting deployment in some locations.

5.4 Rolling Out Smart Meters: Sometimes a Bumpy Ride

Despite some level of resistance due to the reasons detailed here, smart meter rollouts are progressing rapidly. Current projections estimate that by the end of 2014 close to half of all households in Europe, the United States, and Canada will have smart meters installed, and additional installations are planned. Smart meter rollouts have been controversial in some places, while installations have been smooth in others. This variation in ease of deployment highlights variation in utility–community relationships and level and type of trust among communities. The remaining sections of this chapter describe smart meter installations in a few specific regions and communities, including two regions of the United States (California and Massachusetts) and one country in Europe (Germany). These stories of smart meter rollout include tales of both appreciative satisfaction and frustrating mistrust, highlighting the different perspectives and priorities consumers and utilities have with regard to smart meters. While most consumers may not have particularly strong opinions one way or another about smart meter technology, there are people at both ends of a spectrum with deep negative skepticism at one end and passionate positive excitement at the other end.

Grassroots organization to oppose smart meters has been coordinated through multiple mechanisms. In the United States a group called Stop Smart Meters! has emerged to provide coordinated support and knowledge-sharing to local opposition efforts (Stopsmartmeters.org 2014). Smart meter opposition has resulted in some unusual and unpredictable alliances; both conservative Tea Party members and liberal Occupy movement members have become involved in Stop Smart Meters!-type campaigns.

As smart meter technologies are rapidly evolving, maintaining flexibility and integrating future adaptability in technology investments poses a persistent challenge for utilities, regulators, and customers. Utilities are not used to rapidly changing technology – until recently the industry expectation was that the same basic electricity system technology would be used for decades. But with smart meters, there are possibilities for frequent technological improvements – both hardware and software. This creates a major challenge for investment. If a utility invests in one meter design in one year, it will have a difficult time explaining to its customers why they should switch to a different meter in the near future. Another challenge related to the fast pace of technological change is compatibility: it is possible that first-generation meters may not be compatible with second-generation meters if new functionalities are incorporated and industry standards are not agreed upon. This raises issues of standards – how will interoperability of system components be managed? Should a specific type and model of smart meter be mandated, and, if so, by whom?

5.5.1 Smart Meter Deployment in California

Within the United States, the state of California has perhaps the longest and most complicated history of smart meter rollouts. California is an environmental leader, known for advancing innovations for environmental improvement more aggressively than other United States jurisdictions. Within this context, California energy regulators were among the first to approve an electric utility's smart meter initiative, in 2006, with Northern California's Pacific Gas and Electric (PG&E) meter retrofit program. The initial phase of this program proposed retrofitting conventional gas and electric meters with communication devices that would enable monitoring of hourly consumption and variable pricing mechanisms. Responding to rapidly changing smart meter technology, PG&E was successful in amending its proposed meter retrofit program in 2009 to involve installing new meters to replace traditional analog meters (PG&E 2009).

However, the initial rollout was plagued by problems. As this program was implemented, some customers began to experience price spikes in their bills, and they accused PG&E of gouging customers. PG&E's customers were angry; opposition rapidly grew from these negative experiences and forced PG&E to adapt and institute meter opt-out alternatives. During this time additional complaints emerged about PG&E failing to honor customers' smart meter opt-out requests, which exacerbated customer mistrust and frustration with PG&E. So many customers were upset that the California Public Utilities Commission conducted an investigation into the accuracy of the smart meter readings. Installation errors and concerns about faulty signals, overcharging, and health impacts have resulted in dozens of California cities, towns, and counties adopting ordinances to halt smart meter installations (Stopsmartmeters.org 2014). The consumer backlash associated with PG&E's initial rollout of smart meters served as a strong warning to other utilities interested in smart meter deployment and highlighted that smart meters were not a simple technological switch.

To overcome the significant resistance created by the troubled initial smart meter rollout, PG&E has modified their customer service approach to include the opt-out option and also to be more interactive and responsive to customer concerns. Other California utilities have also learned a lot from the PG&E experience. For example, the "Smart Sacramento" project developed by the Sacramento Municipal Utility Department (SMUD) is considered industry "best practice." SMUD worked to educate and engage customers early on in its smart meter program; it used a combination of new communication and messaging strategies in an effort to help customers understand the tangible benefits to them of having a smart meter. SMUD invested in conducting "acceptance testing" of information it was distributing, as well as its customer service (Durand 2014). Messaging focused on "improved service" and "more control over electric bills" and a team of municipal utility employees worked hard to maintain strong and effective communication with a wide range of stakeholders, including SMUD's customers, its employees, elected officials, and the media. Before installation began, SMUD had already engaged in extensive communication, including outreach in five different languages, which was sustained before, during, and after meter installations. In this communication, SMUD prioritized transparency; this included providing information on the success of its smart meter rollout, but also on challenges (Durand 2014). SMUD has become a "best-practice" example of how to effectively manage expectations, develop a community-based engagement plan, and maintain positive relationships with customers through communication with both internal and external stakeholders.

In addition to SMUD, San Diego Gas & Electric (SDG&E) learned from the challenges PG&E faced during its early smart grid rollout and designed a strategy for communication with its customers that was sequenced to provide information and awareness ninety, sixty, and thirty days before installation. With this plan SDG&E was able to set clear expectations among its customers, which resulted in very few complaints or concerns; among the 2.3 million meter installations SDG&E only received about 1,200 complaints (about 0.16 percent), which is less than other programs (SGCC 2013). SDG&E also held contests, offering customers prizes to promote energy conservation measures and engagement in their use of their new smart meters to lower their electricity bill.

Smart meter deployment by Southern California Edison (SCE), the third-largest utility in California, similarly demonstrates learning from PG&E's earlier challenges to engage customers and reduce opposition and resistance. SCE used many different communication tools, including websites, community events, and television and radio ads, which it claims resulted in an early satisfaction rate of 85 percent in its initial smart meter installation process (SGCC 2013).

Despite these innovative communication approaches by the utilities, the level and extent of opposition to smart meters in California has been, and continues to be, strong. Local government responses to these concerns have included attempts to make smart meters illegal in four counties, nine cities, and one tribal community, and resolutions to stop meter installation in more than thirty-two other cities (Hess and Coley 2012). These local attempts have been largely symbolic, because these government entities have no jurisdiction to implement or enforce a ban on smart meters. These local actions did, however, motivate the state's Public Utility Commission to mandate that utilities had to provide an opt-out option; utilities had to offer customers the choice of an analog meter, which had an additional cost to the customer to cover the expenses of not installing a smart meter.

As we seek to understand the controversies surrounding California's smart meter programs, it is important to point out that PG&E was among the first of the utilities in the United States to take on a massive and ambitious rollout of its smart meter program. Unlike other more recent smart meter deployment programs (including National Grid's currently evolving program in Worcester, Massachusetts, described later in the chapter), where a carefully planned pilot strategy has been developed and implemented, PG&E's early attempt included minimal recognition of the potential for public concern, mistrust, and opposition. Reflecting on PG&E's early experience, Jim Meadows, the Director of PG&E's Smart Meter Program, has noted that "originally, people viewed the implementation of the smart grid and the deployment of smart meters as a purely technical change... In actuality, moving toward a smart grid [is] a very substantial transition that requires dialogue and education between the utility and customer base" (Mitchell 2012).

5.5.2 A Smart Meter Pilot Program in Massachusetts

Outside of California, smart meter rollout programs have faced similar types of opposition. But utilities in some other states have also learned from PG&E's difficulties, and have designed carefully planned smart meter pilot programs. Rather than attempting rapid and widespread smart meter deployment throughout entire cities or regions, these pilot projects are slower and have integrated mechanisms for learning. Utilities have previously implemented pilot programs for energy efficiency programs, and now many utilities have ongoing smart grid pilot programs. The details of these smart meter programs demonstrate both the opportunities and the challenges of deploying smart meters, and highlight new types of utility initiatives that take time, resources, and community engagement and that require new and innovative activities for utilities.

One interesting example of a smart grid pilot program is National Grid's "Smart Energy Solutions" pilot project in the city of Worcester, in Central Massachusetts. This program is unique in that it focuses on a mixed socioeconomic population; 15,000 smart meters have been deployed throughout the postindustrial city of Worcester. This project was originally developed in response to Massachusetts legislation, the Green Communities Act, which required each major electricity distribution company operating in the state to coordinate a smart grid pilot program to support the state's grid-modernization efforts. National Grid is an international company that delivers electricity in Massachusetts, New York, and Rhode Island; owns more than 4,000 megawatts of contracted electricity production; and is the largest distributor of natural gas in the Northeast region of the United States (Worcester 2012). One of the primary objectives of Massachusetts' requirement was that the smart grid pilot programs demonstrate advanced smart meters, time-of-use pricing, and at least a 5 percent reduction in peak demand (MA DPU 2012).

National Grid selected the central Massachusetts city of Worcester (population estimated at 181,000 in 2012) for its pilot project, and filed its original pilot program proposal with the Massachusetts Department of Public Utilities (DPU) in April 2009. Worcester was selected for this pilot program for several reasons. First, the city has a diverse population, which will help National Grid learn about smart meters and dynamic pricing for a wide variety of customers. Second, National Grid's distribution network in Worcester is representative of the overhead and underground distribution systems throughout the company's service territory. Third, the city has a number of existing distributed generation project sites and electric vehicle charging stations, which will allow the company to study how these resources affect grid operating conditions; fourth, Worcester has several colleges and universities, which will allow National Grid to offer students experience in the smart grid and renewable energy industries (MA DPU 2012).

The state DPU conditionally approved elements of the pilot program in 2009 but required National Grid to make amendments to its information technology systems; to the marketing, education, and outreach plan; and to its plans for evaluation. In response, National Grid coordinated with the city of Worcester to host a two-day community summit to cultivate community buy-in and engagement, and the pilot was approved in 2011 (MA DPU 2012). The Worcester project is designed to engage both residential and commercial customers in a dynamic pricing program using smart meters and in-home energy management technologies. The pilot also has "grid-facing" components to better manage the distribution network, including the deployment of advanced distribution automation and control, automated distribution system monitoring technologies, fault location devices, and advanced capacitors within the pilot program area. National Grid has also partnered with Clark University to obtain rent-free a Main Street storefront property in which National Grid has created a publicly accessible Sustainability Hub that demonstrates technologies and provides a help desk for pilot participants to learn more about the technologies and the dynamic pricing options.

The Worcester pilot program is an opt-out program, meaning that smart meters will be installed at the households of the pilot participants *unless* the customer requests not to be involved. For participating customers, advanced Intron meters have been installed and the two-year pilot program officially began in spring 2014. The pilot includes assessment of behavior change in four categories of households: (1) households with only the outside-the-home smart meter with web-based access to monitoring, (2) households that also have an in-home display, (3) households with automatic heating, ventilation, and air conditioning (HVAC) controls (no in-home display), and (4) households with in-home display and other advanced controls, including automatic thermostat and HVAC. This $44 million Worcester pilot program, the largest of its kind in Massachusetts, has attracted some significant local controversy and the Stop Smart Meters! organization has been active in the city. Some of those involved in opposing smart meters in Worcester are not city residents, but they passionately engage with and provide support for local opposition. Citizens worried about the health risks of wireless technologies have flocked to Worcester to mobilize an opposition movement (Wright 2013). In spring 2014 a subcommittee of the Worcester City Council voted to instruct National Grid to postpone its pilot project until more was known about the risks of these smart meters.

The Worcester example demonstrates how some of the opposition to smart meters in the Northeast of the United States is part of a larger set of concerns that first emerged in California in 2009 and 2010. In April 2014, a Worcester City Council subcommittee recommended a one-year delay of National Grid's pilot project, citing too many unanswered questions regarding health, security, and privacy (Kotsopoulos 2014).

5.4.3 Smart Meter Deployment in Germany

The story of smart meter deployment in Germany offers some important lessons and insights related to potential downsides of mandating across-the-board deployment of any specific technology. Rather than accepting and working toward the EU Energy Directive requiring that 80 percent of European households have smart meters by 2020, Germany has developed a more gradual and selective approach to installing smart meters, with a focus on deployment in buildings and households where the potential for energy reduction is high. Germany's more selective approach was not received well by the smart meter industry, which has been benefiting from smart meter rollout programs driven by the EU Energy Directive in countries throughout Europe. But the more sophisticated and nuanced approach to smart meter deployment in Germany represents a different prioritization regarding national-level energy system change.

Germany's decision not to adopt a nationwide smart meter rollout mandate was justified with a 2013 cost-benefit report which concluded that in some instances the costs of smart meter deployment outweighed the benefits (Ernst & Young 2013). This report, developed by Ernst & Young on behalf of the German Federal Ministry of Economics and Technology, was used to explain to the EU the German decision to opt out of its commitment to fulfill the EU Energy Directive's requirement of 80 percent deployment by 2020. This report recommends a more gradual and selective smart meter rollout program that takes advantage of existing replacement cycles for meters and does not require smart meters for electricity users with low capacity to reduce their energy use. The case highlights a recurring struggle between the ideal and the realities of smart meters.

One influential smart meter pilot project has been taking place in the city of Karlsruhe, where the municipal utility is partnering with one of the large national-level utilities, Energie Baden-Württemberg AG, on a smart meter pilot project. The local utility has been testing meters with a small group of customers throughout the past few years, but this partnership includes plans for a larger smart meter pilot installation in Fall 2014. One of the unique features of this pilot is the desire to integrate electricity metering with gas, water, and heat.

To understand Germany's cautious approach to smart meters, it is important to consider the larger energy system transformation ongoing in Germany. As part of the national *Energiewende*, an officially adopted energy policy transitioning the country's energy systems away from fossil fuels and nuclear (discussed in more detail in Chapter 6), Germany has strengthened its global leadership role in renewable electricity generation technologies. More than any other country, Germany has invested in social and technical analysis regarding facilitation of an energy system transition. A critical part of implementing the *Energiewende* has included large-scale investments in solar PV; Germany reached over 36 GW of installed capacity of solar PV by

April 2014. This level of investment has been controversial in large part because the substantial solar subsidies are linked to huge increases in consumers' electricity rates. Backlash related to the level of solar investments has resulted in a cautious re-evaluation of all energy system investments, including smart meters.

It is within this context that Germany's resistance to committing to a nationwide smart meter deployment plan has been justified as the state adopts a more graduated plan for smart meter deployment. The German government asserts that the more selective smart meter rollout strategy, rather than mandating that every household have a smart meter by a certain date, is more likely to be effective and have a larger impact. This selective strategy includes prioritizing smart meter installation for certain types of energy users, including those who are particularly high electricity users and those who use combined heat and power.

The German approach highlights the value of allowing smart grid technologies to be selectively deployed in contexts where the technology makes the most sense. This case highlights that regulatory attempts to mandate specific levels of deployment requiring a "one-size-fits-all" approach run the risk of promoting inefficient invest-ments, and also of alienating consumers. If a small household with minimal electricity usage is forced to install a smart meter, and it becomes clear that the meter does not offer much benefit, the household may develop a frustrated or disenchanted view of the potential of smart meters and government programs related to installing these meters. Positive community engagement and public participation are widely recog-nized in Germany as critical to the success of the many implementation strategies associated with the *Energiewende* (Renn et al. 2014), so Germany's selective approach to smart meter deployment has also been influenced by this valuing of the social dynamics of technological change.

5.6 The Future of Smart Metering: Conclusions

Just as smart grid is associated with both exciting promise and potential pitfalls (as reviewed in Chapter 2), so are smart meters. As the two famous quotations at the beginning of this chapter reflect, there are tensions and recurring challenges when the ideal of measuring to manage is juxtaposed with the reality and limits of measurement and data. The smart meter stories in this chapter reflect struggles regarding who has control and who benefits.

Smart meters offer new kinds of control to electricity consumers who manage electricity use and the utilities who manage the flow of electricity throughout the system. The new level of control for utilities comes with a new level of skepticism among some consumers about how the utility will use its control. Smart meters also offer new kinds of cost and efficiency benefits to both consumers and utilities through the enhanced capacity to monitor and measure electricity use. Skepticism has emerged here too about whether the economic benefits of smart meters are greater for the utilities or for the consumers. The strong opposition to smart meters that is felt deeply by some reflects concern about a loss of control associated with a mistrust of governments' and utilities' protection of privacy and health.

In the United States, the promises of smart meters relate primarily to reducing peak load by providing information to electricity users about their use and, to a lesser extent, using time-of-use pricing mechanisms. Smart meters can also offer resilience in terms of monitoring consumption and by directly controlling demand through preset controls. In Europe, despite a top-down EU directive encouraging smart meter installation, there has been wide variation in member state responses. Another major promise of smart meters is that they will save money and reduce costs to consumers, utilities, and society. Many customers have been skeptical of this promise because some electricity bills actually increased immediately after smart meters were installed. Another important promise of smart meters is the environmental improvement associated with enhancing efficiency, enabling changes in electricity use patterns, and reducing overall electricity demand.

Different actors (Chapter 4) have different priorities for smart meters. For some smart meters are a critical tool to manage energy; for others smart meters feel like an imposition with no tangible benefits. The passionate opposition to smart meters can be understood, at least partially, as a symptom of deeper mistrust of government and industry in the United States. As Americans react to revelations of NSA spying and corporate data breaches, mistrust of government and industry grows, but there are very few venues for expressing outrage or standing up against increased monitoring. Smart meter installations therefore provide one specific venue for citizens to question the benefits of collecting all of this electricity usage data.

Opposition and controversy are expected parts of any technology development. At least in the case of smart meters, it seems clear that opposition is not solely about the technology in question. Social and political concerns such as choice and privacy may be at least as important as the technology itself. When people are required to make changes without feeling like they have had a choice, concerns and mistrust emerge even more strongly. If, on the other hand, people have opportunities to learn about new technologies before being presented with an *a priori* decision, they are more likely to be open to the possibility of change.

The examples provided in this chapter demonstrate the complex nature of smart meter deployment. Different local contexts are shaping patterns of support or opposition. The accumulated experience of smart meter deployment continues to grow as more and more smart meters are installed and the novel technology becomes less threatening as it becomes mundane. This wealth of experience is shared among utilities and communities and activist groups. In many places, including Texas and Illinois in the United States, smart meters have been installed with virtually no apparent citizen concern. Canada provides an interesting national-level example of this variation (Mallett et al. 2014). While smart meter deployment has been relatively smooth in Ontario (some resistance has emerged due to concern about unfair pricing), in British Columbia tensions have emerged as marijuana growers, some of whom have been stealing electricity for decades, have been threatened by smart meter installation. Quebec has also experienced strong opposition to smart meters, in part due to mistrust of the state-owned Hydro-Quebec and the loss of meter-reading jobs (Mallett et al. 2014).

Another tension fueling smart meter power struggles relates to the perspectives on whether or not these devices provide more or less autonomy to individual electricity consumers. While the smart meter can offer tools to help customers change their electricity use, it can also be viewed as a tool that takes control away from consumers and shifts the balance of power further to utilities, which may remotely monitor energy use and directly control some appliances. This tension reflects the delicate balance between privacy and system function.

Although the smart meter has become, for some, synonymous with smart grid, the meter is only one small (but critical) part of the electricity system. The disproportionate attention being paid to smart meters was described well by one of the energy system experts we talked to, who said: "[A] smart meter without the rest of the smart grid is like having an iPad without internet."

Key considerations for understanding the future of smart meters include the degree to which smart meter deployment is aligned with individual and collective goals. Another consideration is the balance between perceived short-term benefits and long-term benefits, and perceived benefits for utilities versus perceived benefits for electricity users.

References

ACLU. (2014) Time to Rein in the Surveillance State. www.aclu.org/time-rein-surveillance-state-0

American Cancer Society. (2012) Smart Meters. http://www.berginsight.com/ReportPDF/ProductSheet/bi-sm9-ps.pdf

BioInitiative Working Group. (2012) BioInitiative 2012: A Rationale for Biologically-based Exposure Standards for Low-Intensity Electromagnetic Radiation. www.bioinitiative.org/table-of-contents/

Berg Insight. (2013) Smart Metering in Europe. M2M Research Series. www.berginsight.com/ReportPDF/ProductSheet/bi-sm9-ps.pdf

Berst, J. (2013) Nein! German ministry rejects smart meters. *Smart Grid News.com.* http://www.smartgridnews.com/artman/publish/Technologies_Metering/Nein-German-ministry-rejects-smart-meters-5958.html#.Uv4rOYXwr9o

Bomkamp, S. (2013, July 9) GE to Make Smart Meters in Chicago. *Chicago Tribune.* articles.chicagotribune.com/2013-07-09/business/chi-chicago-smart-meters-20130709_1_smart-meters-new-meters-comed

Brown, H. S. (2014) The Next Generation of Research on Sustainable Consumption. *Sustainability: Science, Practice, & Policy*, 10(1), 1–3. sspp.proquest.com/archives/vol10iss/editorial.brown.html.

California Council on Science & Technology. (2011) *Health Impacts of Radio Frequency from Smart Meters Response.* www.ccst.us/publications/2011/2011smart-final.pdf

Clarke, C. (2012) Are California Smart Meters Causing Fires. *KCET Rewire.* www.kcet.org/ news/rewire/technology/are-california-smart-meters-causing-fires.html

Darby, S. 2006. The Effectiveness of Feedback on Energy Consumption: A Review for Defra of the Literature on Metering, Billing and Direct Displays. www.eci.ox.ac.uk/research/ energy/downloads/smart-metering-report.pdf

Darby, S. (2010) Smart Metering: What Potential for Householder Engagement? *Building Research and Information,* 38, 442–457.

DoE. (2010) Budget Information. www.oe.energy.gov/budget.htm

Durand, P. (2014) Munis Leverage AMI lessons. *Intelligent Utility.* www.intelligentutility. com/article/14/03/munis-leverage-ami-lessons

Electric Light & Power. (2014) Landis+Gyr Rolls Out Smart Meters to 98 Percent of Finland Homes. www.elp.com/articles/2014/02/landis-gyr-rolls-out-smart-meters-to-98-percent-of-finland-homes.html

EMF Safety Network. (2014) EMF Safety Network Website. http://emfsafetynetwork.org/

EPRI (2010) An Investigation of Radiofrequency Fields Associated with the Itron Smart Meter. *Electric Power Research Institute,* 1–222.

Ernst & Young. (2013) Cost-benefit Analysis for the Comprehensive Use of Smart Metering. German Federal Ministry of Economics and Technology. www.bmwi.de/English/ Redaktion/Pdf/cost-benefit-analysis-for-the-comprehensive-use-of-smart-metering-systems,property=pdf,bereich=bmwi2012,sprache=en,rwb=true.pdf

Eurelectric. (2013) Utilities: Powerhouses of Innovation. www.eurelectric.org/media/79178/ utilties_powerhouse_of_innovation_full_report_final-2013-104-0001-01-e.pdf

FERC. (2013) Assessment of Demand Response & Advanced Metering. www.ferc.gov/legal/ staff-reports/2013/oct-demand-response.pdf

Goldsworthy, A. (2007) The Dangers of Electromagnetic Smog. www.hese-project.org/hese-uk/en/papers/electrosmog_dangers.pdf

Greentech Media. (2014) www.greentechmedia.com/articles/read/smart-meter-penetration

Hess, D. J. (2013) Smart Meters and Public Acceptance: Comparative Analysis and Design Implications. Paper presented at the meeting of the Sustainable Consumption Research and Action Network, Clark University, June 12–14, Worcester, Massachusetts.

Hess, D. J. (2014) Smart Meters and Public Acceptance: Comparative Analysis and Governance Implications. *Health, Risk & Society,* 14(3), 243–258.

Hess, D. J. and J. S. Coley. (2012) Wireless Smart Meters and Public Acceptance: The Environment, Limited Choices, and Precautionary Politics. *Public Understanding of Science.* DOI:10.1177/0963662512464936

Johnston, P. (2013, August 26) Smart Meters: Good Idea or a Lot of Hot Air? *The Telegraph.* www.telegraph.co.uk/finance/newsbysector/energy/10267013/Smart-meters-good-idea-or-a-lot-of-hot-air.html

JRC Scientific and Policy Reports. (2013) *Smart Grid Projects in Europe: Lessons Learned and Current Developments.* Brussels: European Commission.

Keller, K. H. (2008) From Here to There in Information Technology: The Complexities of Innovation. *American Behavioral Scientist,* 52, 97.

Kotsopoulos, N. (2014, April 24) Worcester City Council Panel Wants 'Smart Grid' Delay. *Worcester T&G.* http://www.telegram.com/article/20140423/NEWS/304239484/1116

Krishnamurti, T., D. Schwartz, A. Davis, B. Fischhoff, W. Bruine de Bruin, L. Lave, and J. Wang. (2012) Preparing for Smart Grid Technologies: A Behavioral Decision Research Approach to Understanding Consumer Expectations about Smart Meters. *Energy Policy,* 41, 790–797.

Langheim, R. (2013) Smart Grid in the US: Visioning and Framing Opportunities for Electricity System Change. In *Environmental Science and Policy Program, Department of International Development, Community and Environment*. Worcester, MA: Clark University, 1–56.

M2M Research. (2013) Smart Metering in Europe. www.berginsight.com/ReportPDF/Summary/bi-sm10-sum.pdf

MA DPU. (2012) Petition of Massachusetts Electric Company and Nantucket Electric Company, each d/b/a National Grid for Approval of a Smart Grid Pilot Program. http://www.env.state.ma.us/dpu/docs/electric/11-129/8312dpuord.pdf

Mallett, A., R. Reiber, D. Rosenbloom, X. D. Philion, and M. Jegen. (2014) *When Push Comes to Shove: Canadian Smart Grids Experiences Through the Media*. Paper presented at 2014 Canadian Political Science Association Annual Conference. Brock University, May 27–29, 2014.

Mitchell, T. (2012) PG&E Shares Smart Meter Lessons. *Fierce Energy*. www.fierceenergy.com/story/pge-shares-smart-meter-lessons/2012-08-01

Navigant Research. (2014) Home Energy Management: New Players, Technology Update, and Market Outlook – webinar. www.navigantresearch.com/webinar/home-energy-management-2

Paetz, A.-G., E. Dutschke, and W. Fichtner. (2011) Smart Homes as a Means to Sustainable Energy Consumption: A Study of Consumer Perceptions. *Journal of Consumer Policy*, 35, 1, 23–41.

PG&E. (2009) Application to the Public Utility Commission of California of Pacific Gas and Electric Company for Authority to Increase Revenue Requirements to Recover the Costs to Upgrade its SmartMeter™ Program (U 39 E). www.pge.com/includes/docs/pdfs/shared/customerservice/meter/smartmeter/cpucdoc_smartmeterprogram-upgrade.pdf

Pidgeon, N., R. Kasperson, and P. Slovic. (2003) *The Social Amplification of Risk*. Cambridge, UK: Cambridge University Press.

Renn, O., W. Köck, P.-J. Schweizer, J. Bovet, C. Benighaus, O. Scheel, and R. Schröter. (2014) Public Participation for Planning New Facilities in the Context of the German "Energiewende". Policy Brief Edition 01/2014. Helmholtz Association Alliance ENERGY-TRANS.

Rivaldo, A. (2012) Report on Health and Radiofrequency Electromagnetic Fields from Advanced Meters. *Public Utility Commissions of Texas*, 1–86.

Smart Grid Consumer Collaborative. (2013) 2013 State of the Consumer Report. http://smartgridcc.org/wp-content/uploads/2013/01/SoCR-2013_1.24.pdf

Shove, E., M. Pantzar, and M. Watson. (2012) *The Dynamics of Social Practice: Everyday Life and How it Changes*. London: Sage.

Slovic, P. (2006) *The Perception of Risk*. London: Earthscan.

Slovik, P. (2010) *The Feeling of Risk*. London: Earthscan.

Stopsmartmeters.org. (2014) Stop Smart Meters! http://stopsmartmeters.org/

Synerge Worcester. (2012) Synerge Worcester. In *National Grid: Founding Strategic Partner*. http://synergeworcester.com/team/national-grid/

Weiss, M., F. Mattern, T. Graml, T. Staake, and E. Fleisch. (2009) Handy Feedback: Connecting Smart Meters with Mobile Phones. www.vs.inf.ethz.ch/publ/papers/weismark-handyFeedback-2009.pdf

WHO. (2006) Electromagnetic Fields and Public Health. www.who.int/peh-emf/publications/facts/fs304/en/index.html

World Health Organization. (2014) Electromagnetic Fields (EMF) Standards and Guidelines. www.who.int/peh-emf/standards/en/

Wright, Z. (2013) Will Smart Meters Be Stopped in Worcester? *GoLocal Worcester*. www.golocalworcester.com/news/will-smart-meters-be-stopped-in-worcester

6

Wind on the Wires[1]

6.1 Tensions and Synergies of Large-scale Wind and Smart Grid

For many, one of the most appealing and valuable promises of a smarter grid is the ability to integrate more renewable energy into the electric system. More renewables allow for lower-carbon electricity generation, facilitate a reduction in fossil fuel reliance, and enable diversification of sources of electricity generation. As climate and renewable energy advocates are calling for a transition to low-carbon electricity systems, it is widely recognized that wind, solar, and other renewable power has great potential to contribute a greater share of total electricity generation. In this chapter we focus on large-scale wind development and explore how it has both required and influenced the simultaneous development of smart grid.

To integrate wind into the electric system, smart grid has come to represent a crucial set of technologies that will continue to ensure grid reliability and resilience while allowing wind to play a larger role in the electric power grid. Some estimate that wind could ultimately provide between 20 and 50 percent of total global electricity generation (NREL 2008; Xu et al. 2009; Meegahapola and Flynn 2010). Scaling up wind power to these levels requires both technical and social innovation and involves installing wind turbines, new transmission networks, and new operations to integrate wind. The rapid expansion of wind power has led to an evolving shift from focusing on turbines to an expanded systemwide perspective to support wind power integration.

The recent history of wind power development illustrates how policies to promote renewable generation are changing electric grid operations and shaping smart grid development; at the same time, policies to promote grid modernization are changing the development and operation of renewable energy technologies. In just over twenty years, this dynamic coevolution has transformed wind power from a boutique

---

[1] Wind on the Wires is the name of a Minnesota-based nonprofit organization whose mission is to overcome barriers of bringing wind energy to market: see http://windonthewires.org/

experiment to a major source of power in many places. These changes have shifted the power dynamics among system actors and changed institutions. The simultaneous development of smart grid and renewable energy has also created new challenges and struggles. Large-scale wind power has changed transmission planning and financing, shifted the operation of the electric power grid, and shifted the economics of the electric system. These shifts in planning, siting, and financing of long-distance transmission lines to connect distant large-scale wind generation with electricity demand have created multiple struggles as new challenges and opportunities have emerged for electricity system actors. The synergies and tensions associated with smart grid and large-scale wind development vary regionally, which demonstrates the importance of the social and political contexts shaping electricity system change.

The uneven spatial patchwork of wind deployment is due to multiple factors: a lack of wind resources is limiting in some places, while political support has been insufficient to stimulate development of large-scale wind systems in other systems. In locations with large-scale deployment the integration of wind is also affecting the operation and value of existing generation, changing electric sector economics, and undermining traditional electric utility business models. As electricity system incumbents are forced to adapt – sometimes at great cost – political support for renewable energy is changing rapidly. These shifting social and political dynamics of renewable energy development take on different forms in different contexts.

In this chapter, we explore the interactions between smart grid development and large-scale wind power deployment and examine their impacts on several energy system transitions. These interactions are shaped by multiple institutions and policies at the local, state, regional, national, and sometimes international levels. To highlight the multiple factors influencing integrated development of large-scale wind power and a smarter grid, we develop three case studies. The first case focuses on wind development in the state of Texas, the U.S. state with the most installed wind power. Strong wind resources, a history of wind power use for water and oil pumping, and favorable electricity system economics in Texas have supported the development of over 10,000 MW of wind power. The second case focuses on relationships between regional development and wind power in the Upper Midwest region of the United States. We explore how states and the regional electricity system operators have worked together to plan for and integrate wind power into the electric system. With more than 13,000 MW regionally and 25,000 MW planned in the Midwest, wind power development has required unprecedented levels of regional coordination and cooperation. The third case focuses on national-level wind development in Germany, a country in the midst of a nationally planned energy transition, the "*Energiewende*," also discussed in Chapter 5. With more than 33,000 MW of installed wind power and 36,000 MW of solar PV, Germany has become a global leader in the development and integration of renewables. Together, these cases allow us to explore more deeply the relationships between electricity system change and scaling up renewables. As we describe each case, we focus on how the coevolution of wind development and a smarter grid are integrating technological advances and social changes in laws and markets and associated regulatory, financial, and legislative institutions.

Within each case we explore variation of and linkages between actors, technologies, and institutions at multiple decision nodes. Who controls wind power development and who benefits differs across the scenarios and over time. In this chapter we focus on the critical interactions and co-development of large-scale wind power and smart grid to highlight evolving creative tensions of smart grid development embedded in larger system-wide change. We also explore how polices used to promote wind power are shaping smart grid development. The Texas case highlights how policies to encourage renewable electricity generation have led to the development of new transmission lines and also changed electric grid operations. In the Upper Midwest and Germany, transmission lines to connect remote wind sites to load centers often cross multiple jurisdictions, disrupting traditional siting and cost allocation practices. These case studies explore the shifting alliances and tensions across all three systems. In this chapter we explore the question of who benefits and who looses when large-scale wind power development and smart grid are linked.

We begin the chapter by reviewing the potential of large-scale wind power, including a brief historical perspective on the coevolution of wind power technology and electricity system development beginning in the 1800s. This is followed by the Texas, Upper Midwest, and German case studies. We conclude the chapter by discussing the commonalities among the three cases.

6.2 Wind Power in Context

Globally, wind power is increasing rapidly; with over 310,000 MW of installed wind capacity by the end of 2013, wind power has become the fastest growing energy technology. This rapid growth in wind power highlights the transformation of the technology from a boutique energy source to a critical technology shaping grid management and electricity markets.

Wind is created from the rotation of the earth, the shape of the earth surface, and the uneven heating of the earth's atmosphere. These factors combine to form different wind patterns across the earth that can be harnessed to push turbine blades, which then spin the rotors driving a shaft to generate electricity. Wind power is a "variable resource," only producing electricity when the wind is blowing. Wind turbines generate electricity above a minimum cut-in speed, approximately 7–10 miles per hour. Most turbines produce power at full capacity at speeds of 25–30 miles per hour (the rated speed) and cut out if the wind speed is above 45–80 miles per hour to protect the turbines if the wind is too strong. Wind turbine technology has been improving to expand the cut-in and cut-out operating range which increases wind power's capacity factor, or percentage of time the turbine is operating at its rated capacity. Wind turbine capacity factors vary because they depend on both the technology and the site-specific wind resource. Wind power capacity factors have been steadily increasing, from less than 18 percent in the mid-1990's to current levels ranging from 23 to 36 percent (Wiser and Bolinger 2013; Kaldellis and Zafirakis 2011). The variability of wind power is altering how the electric grid is managed.

Requirements for integrating wind into the grid have changed over the past decade. When there were only small amounts of wind power generating electricity, special operational protocols were not needed. Conventional wisdom assumes large-scale wind power requires other electricity-generating resources to be ready to quickly come online due to unpredictability regarding when the wind might stop blowing. However, development of a smarter grid, accompanied by better meteorological forecasts, has made grid operators more adept at integrating wind into the electric system. While system operators originally had to call wind plant operators to curtail turbine output when transmission congestion or system constraints required, with improved communication technology this process has become automated in most systems. In some electricity systems, wind plants now bid into day-ahead electricity markets and are scheduled like conventional generation resources. A smarter grid for wind means new market mechanisms, new systemwide data integration, new control algorithms, and the incorporation of detailed wind prediction models, enabling grid operators to integrate greater levels of wind power. These systems continue to evolve, and engineers are currently developing additional controls to allow wind plants to be more flexible and better integrated into the electric grid.

Throughout human history, people have harnessed the wind to meet multiple societal functions, from milling grain to pumping water and oil, and now creating electricity. Each use of wind power involves evolving relationships among people and technology and each use has fulfilled the needs of different actors. The current evolution and integration of wind into the electric system is but another chapter in the long use of wind power to meet changing human desires.

6.2.1 The Evolving History of Wind Technologies

Wind power was used to grind grain in Persia in the seventh century and in Europe and China from the 1100's (Musgrove 2010; U.S. Department of Energy 2011). Throughout five centuries in Northern Europe, wind power was harnessed by windmills to grind grain and became a dominant technology, driven by feudal economics. In the feudal society, the windmill was owned by the lord of the manor and this was where tenants were required to grind the grain they grew. In the Netherlands, for example, one windmill was able to grind enough grain to feed 2,000 people. In addition to grinding grain, windmills pumped water and helped to drain the Rhine Valley marshes. Windmills also sawed wood and crushed oil seeds. Windmills dominated the European landscape for 500 years, but their use declined rapidly with the advent of the steam engine in the late 1700's and the import of grain from North America. In North America in the 1850's, wind pumps became a key technology which enabled European settlers to move westward, using the pumps to draw water from deep underground aquifers for farming and ranching (Galbraith and Price 2013). In the mid-1800's, wind pumps also became a crucial transportation technology. The railways used larger wind-driven pumps to store water along the railway tracks for steam engines (Musgrove 2010).

6.2.2 Creating Electricity from Wind

Unlike grinding grain or pumping water, both of which require a slow, steadily turning rotor, generating electricity from wind requires a rapidly spinning rotor to drive the generator. In the late 1880's wind power was first used to generate electricity almost simultaneously in Europe and the United States. In 1887 James Blyth, a university professor from Glasgow, Scotland set up a small turbine near his holiday cottage (Price 2013); in 1888, electricity pioneer Charles Brush installed the first turbine to power the lights in his mansion in Cleveland, Ohio, in the United States (Scientific American 1890). While these two inventors did not appear to know each other, the use of windmills to generate electricity in isolated locations spread. Windmills connected to generators were used to power lighthouses in France and many rural farms in North America, for example. However, generating electricity from wind posed several technical challenges.

For one, managing variable wind speed proved difficult and the battery storage banks used by these early wind turbines to store electricity often failed. Danish wind turbine advances helped to overcome challenges in managing variable wind speeds. Spurred by the energy shortages in World Wars I and II, Danish engineers worked to improve turbine function through better gear box design and controls.

In the 1950's and 1960's, low global energy prices stalled further development of wind turbines, although pollution from coal-fired power plants spurred some cursory interest in alternative energy and wind development in the 1960s. It was not until after the 1973 Oil Crisis and the accompanying fourfold increase in the price of oil that investment in alternative energy technology development significantly increased (Musgrove 2010).

In the United States, Denmark, and Germany, large national-level investments supported wind power research. Wind turbine development benefited from parallel advancements in materials, aerodynamics, and structural engineering. As is the case with all technology research and development, not all experiments were successful, but research teams made significant advancements in turbine design (Heymann 1998). From 1975 to 1988 the United States spent $427.4 million on wind power research and development (R&D) to improve turbine technology, Germany invested $103.3 million, and Denmark supported wind research with $19.1 million (Heymann 1998). Much of the German and American research was focused on designing effective and efficient larger wind turbines that would be able to generate larger amounts of power than the existing turbines of the time. The Danish program took a different, bottom-up approach (Vestergaard, Brandstrup, and Goddard 2004).

During the late 1970's and early 1980's, the U.S. program focused primarily on large-scale turbine development. Led by the National Aeronautics and Space Administration (NASA) and involving large aerospace and energy firms such as Boeing, McDonnell Douglas, Lockheed, and Westinghouse, the U.S. program developed four machines; however, they suffered from mechanical difficulties, with the most successful running for 8,000 hours (Heymann 1998). From 1987 to 1992 NASA's MOD-5-B operated for a total of 20,000 hours in Hawaii before it was dismantled due to chronic mechanical failures and high costs. After this, the federal United States wind turbine development program was ramped down (Heymann 1998).

In Germany, the Ministry for Research and Technology *(Bundesministerium fur Forschung und Technologie,* BMFT) began to invest in research to develop large-scale wind turbines in 1974. Like the U.S. program, Germany's was also focused primarily on the development of large machines. An enormous 3MW two-bladed Growian (or big wind power plant) installed from 1983 to 1987 became a very public failure, as it was dismantled due to design failures five years after construction, with only 420 hours of operation (Musgrove 2010). Smaller prototypes like the 370kW mono-blade machine, Monopteros, had more success, but were noisy and not commercially successful; the program ended in the early 1990's (Heymann 1998).

In contrast to the large national programs pursued by the United States and Germany, Denmark's wind development strategy focused on developing a smaller-scale or "market pull" wind program (Vestergaard et al. 2004). This program used engineers and artisans focused on developing smaller, commercially successful grid-connected wind turbines for the Danish market. As early prototypes were successful, they gradually increased the turbine size, with their designs soon dominating early grid-connected wind projects.

6.3 Wind in the United States

To provide context for the Texas and Upper Midwest U.S. case studies, this section outlines the policy context for wind power development in the United States. The United States has excellent on and offshore wind resources; recent studies estimate a potential 10,000–12,000 GW (thousand MW) of rated capacity, with a gross capacity factor of over 30 percent (NREL and AWS Truepower 2014), which includes estimations of over 4,000 GW of offshore wind (Schwartz et al. 2010). The entire U.S. electric grid installed capacity is a little less than 1,000 GW, so theoretically at least, the United States could meet all of its current demand with a massive deployment of wind power. The 61,000 MW of currently installed wind turbines are not evenly spread across the country, but clustered in places that have both supportive policies and strong wind resources: Texas (12,214 MW), California (5,587 MW), Iowa (5,133 MW), Illinois (3,568 MW), and Oregon (3,153 MW) are the states with the most installed wind capacity (NREL 2013). In the Upper Midwest, a total of 13,000 MW of wind power capacity is online.

6.3.1 United States Policy Context for Wind Development

Just as the history of United States wind turbine research and development was linked with the geopolitics of energy, changes in renewable-related state and federal policies were also linked to the larger global energy landscape. In late 1973, the OPEC Oil Embargo rapidly quadrupled the price of oil, and the subsequent gasoline shortages profoundly shifted U.S. attitudes toward energy for the next decade. When President Carter signed the 1978 Public Utilities Regulatory Policy Act (or PURPA), Section 210 opened the utility industry to third-party producers, including wind energy developers. Section 210 of this bill required electric utilities to purchase electricity from qualifying facilities at a price equal to the utility's avoided cost of generation (Hirsh 1999). Although the significance of this requirement was not broadly recognized at the time, this national-level policy had a large impact on wind development.

In addition to the federal policies shaping the energy landscape, U.S. state policies also have influenced wind development. States retain statutory authority to approve and site projects, set electric rates, and implement multiple types of energy policy and incentives. It was up to individual state Public Utilities Commissions (PUCs) to set the "avoided cost" rate required by PURPA. In California, the PUC stipulated that qualifying facilities could be paid 0.07 $/kWh for every kilowatt sold. Also in 1978, California passed legislation to give tax credits for solar and wind energy development. With these state and federal policies, coupled with earlier federal incentives which granted tax credits for capital investments and energy sector investments, the economic incentives for developing renewable energy generation projects became enticing.

California, a state that has always been progressive in environmental and energy programs, soon became a leader in early wind energy development in the United States. The combined state and federal policies amounted to a 50 percent tax credit for wind and solar and, with PURPA guaranteeing market access, these policies created the first U.S. wind boom in California (Musgrove 2010).

Throughout the 1980's the first wind farms were installed, with roughly 4,000 grid-connected 100 kW wind turbines installed at Altamont Pass, 60km east of San Francisco. Other wind farms were created in San Gorgonio Pass near Palm Springs and the Tehachapi Mountains near Bakersfield. From 1981 to 1985, more than 12,000 wind turbines were built in California, with a total installed capacity of over 1,000 MW. These early wind farms did not follow a standard design: while most were horizontal axis machines, there were more than thirty different turbine configurations, and 500 vertical axis machines were also installed. Unfortunately, like the U.S. wind turbine development program, the new wind fleet was plagued with performance issues. While California and federal policies supported the *installation* of wind capacity, the policies did not offer direct support for electricity production. While some of the machines proved reliable, many of the turbines installed by smaller companies were untested, lightweight "proof of concept" machines that could not stand the rigors of real-world operation. Failures were common and the nascent wind sector lost ground and public confidence suffered. To recover, Danish turbines were imported, selling for three to six times more in California than in Denmark. When the U.S. federal tax incentives expired in 1985, Danish turbines dominated the California wind market. By 1990, the first United States wind boom was over, and of the 1,820 MW of wind power that had been installed, only 1,500 MW was operational. Although poor performance as well as public concern about wind farm impacts on birds and viewsheds damaged the reputation of the wind industry, large-scale grid-connected wind was now a reality.

During the California boom in the 1980s, turbine technology evolved considerably. Refinements in turbine design allowed for bigger rotor diameters, higher hub heights, and more power output per turbine. Advanced electric power converters allowed for variable rotor speed to increase power output. However, after the boom, the lack of new supporting policies meant that innovation in wind turbines shifted from the United States and back to Europe for the next decade.

With the Energy Policy Act of 1992 (P.L. 102–486), the federal government passed the Production Tax Credit (PTC), which provided $0.015 for every kilowatt hour of electricity generated from wind power. Unlike the earlier California law which paid for installed capacity, this corporate tax credit was structured to compensate electricity generation from wind. The tax credit was too low to incentivize installation, so throughout the 1990s wind plant installations were minimal. In the 2000's the widespread adoption of state-level Renewable Portfolio Standards (RPS) spurred the next phase of growth in wind power. The most rapid growth in wind occurred in 2008, 2009, and 2012 (Figures 6.1 and 6.2).

By 2013, wind power was generating up to 25 percent of in-state power in places such as Iowa and South Dakota, but the lack of transmission lines to move the power

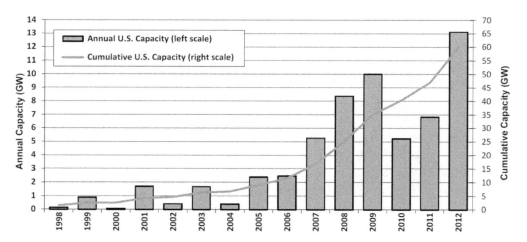

Figure 6.1 Annual and cumulative installed wind power capacity. The production tax credit expired in 2000, 2002, and 2004, shaping installed wind capacity. Threats of PTC expiration in 2013 helped to drive record capacity installations in 2012. Source: Wiser and Bolinger 2013

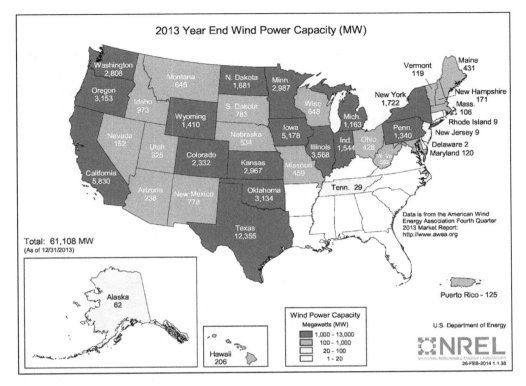

Figure 6.2 Installed wind power capacity in the United States at the end of 2013.
Source: NREL 2014

to demand centers also led to wind power being curtailed or shut down due to
transmission constraints. This pattern of curtailment shifted the focus from building
wind turbines to enhancing the transmission system to support them.

6.3.2 From Turbines to Transmission

While sophisticated turbine designs are crucial for maximizing power production,
wind turbines only help the electric system if they are connected to the grid. The rapid
development of large-scale wind power in the United States starting in 2007 has
underscored the need for additional transmission-line expansion. As transmission
line-siting authority rests with state PUCs, creating coordinated transmission plans
to facilitate wind development has been an ongoing challenge (Vajjhala and Fischbeck 2007, Klass and Wilson 2012). Congress attempted to address this issue in the
Energy Policy Act of 2005, by giving the Federal Energy Regulatory Commission
(FERC) and the Department of Energy (DoE) the authority to identify and establish
National Interest Electric Transmission Corridors and establish fast-track siting procedures. This approach, however, has not been successful; while the DoE identified
corridors in the Southwest and Mid-Atlantic, environmental groups challenged

FERC's "fast-track" authority on the grounds that it was bypassing state jurisdiction, infringing on property rights, and overriding important environmental laws. In February 2011, the Ninth Circuit Court of Appeals ruled in favor of the environmental groups, effectively stalling the attempt to streamline transmission-line expansion.

Interstate transmission planning has always been challenging in the United States and efforts have devolved from a coordinated federal role toward encouraging Regional Transmission Organizations (RTOs) or transmission coordinating councils to create regional transmission plans in their service territories. FERC Order 1000 was the latest attempt to encourage multistate coordination for transmission planning by addressing cost allocation issues. This FERC order emphasized regional transmission planning to help meet state public policy objectives such as state RPS or energy efficiency standards, in addition to helping to achieve the traditional economic and reliability considerations that have been the foundation of future grid-planning efforts (FERC 2011).

6.3.3 From Transmission to Integration

While there are transmission challenges for wind development in many parts of the country, federal efforts to help integrate wind power into the electric system operation and power markets have also been crucial. In 2005, FERC adopted Orders 661 and 661A, which established technical standards and procedures for public utilities connecting large wind projects (FERC 2005, Porter et al. 2009). These standards were established to ensure reliability as wind grew to be a more important energy source, and include the low-voltage ride through, which requires wind generators to remain connected to the grid for a certain time during system faults and low-voltage events. These standards also included the power factor design criteria for reactive power and wind turbines. This order also required wind plants to have SCADA capability and be able to receive instructions – but transmission operators were not authorized to control wind plants. Additionally, the wind developer was able to satisfy the interconnection request and enter the transmission queue with preliminary design specifications, with the agreement to provide detailed design information six months later. Regional Transmission Organizations (RTOs) were required to adopt these orders to overcome some of the barriers faced by wind developers. The experiences of ERCOT (Electricity Reliability Council of Texas which is not regulated by FERC) and the Midcontinent Independent System Operator (MISO) region are further explored later in the chapter.

Building on this context, the next part of this chapter explores in more detail three case studies: (1) Texas, the U.S. state with the most installed wind power; (2) the Upper Midwest of the United States, where states and the electric grid system operator have worked together to plan for and integrate wind power into the electric system; and (3) Germany, a nation in the midst of an energy transition, the *Energiewende*, and a leader in the development of large-scale wind. Together, these cases allow us to more deeply explore the struggles and synergies involved in the coevolution of large-scale wind power and smart grid.

6.4 Texas: Wind and Transmission

Texas leads the United States in installed wind capacity, with more than 12,000 MW of wind on the system. The scale of wind development in Texas has been made possible by the co-development of a sophisticated, flexible, and extensive grid system that has involved multiple actors. Coordinated action from grid operators, wind developers, the state legislature, the PUC, and transmission developers has resulted in rapid and responsive changes to the electric system. Wind development has been made possible with state-supported transmission system upgrades, integration of sophisticated weather monitoring systems, and approaches to integrate wind resources into competitive electricity markets.

6.4.1 Building Wind Power in Texas

The Texas wind case highlights the state's long history of wind use and the recent context for rapid growth of wind power. Wind development was at times threatened by insufficient transmission capacity, but a concerted effort to address this issue from the regional transmission organization, ERCOT, the Public Utility Commission of Texas (PUCT) and the energy community led to novel system planning and operation to accommodate the new demands that wind made on the system.

In Texas, wind power has played a pivotal role in state history, especially in the dry western portion of the state (Galbraith and Price 2013). Wind power was first a crucial component of the water system and early transportation systems which opened up the state to ranchers in the 1800's. The dry Texas Panhandle receives less than eighteen inches (46 cm) of rainfall per year, making tapping groundwater critical for the early railroads, for raising stock, and for providing water to supply cities. By the 1880's, large windmills called Eclipses (5.4m in diameter) and water tanks were located every thirty miles along the railway tracks to service steam locomotives. Texas became the largest windmill state in the country, with ranchers installing windmills to pump water for livestock, and town residents built turbines to provide drinking water. Midland Texas, now famous as the center of oil production, was known as "Windmill Town," as almost every house had a wooden tower with a spinning rotor on its property, and by the early 1900's roughly 50 percent of all windmills produced were sold to Texans (Galbraith and Price 2013). In a recent book about the wind boom in Texas, Galbraith and Price write: "On the desolate plains, the windmill had become a beacon of civilization" (p. 16). The powerful three million-acre (12,000 km^2) XIT ranch installed more than 300 windmills and King Ranch more than 200 windmills which operated until the 1960's (Galbraith and Price 2013). Rural electrification in the 1930's allowed West Texas ranchers to replace windmills with electric pumps, and the dominant panorama of cattle clustered around a spinning windmill faded. In addition to water, Texas' windmills were also used to pump Texan oil and played an important part in early industrial development. As electricity became widely accessible, windmills for pumping water and oil were replaced by more easily maintained and less costly electric pumps.

Texans began making electricity from wind in the early 1980's. Texas has strong wind resources; an ERCOT study estimates that potential wind resources in-state are more than 100 GW of wind with a 35 percent capacity factor and 35 GW of wind with a 40 percent capacity factor (Lasher 2008).

Texans have tinkered with windmills since the late 1800's on the ranch and they have actively participated in the development of the new generation of wind turbines and the rapid growth in wind power. In 1981, Michael Osborne installed five turbines in Pampa, Texas, making it the second wind farm in the United States (the first was in southern New Hampshire; Galbraith and Price 2013). Federal policies provided crucial incentives to develop the first Texas wind farm; PURPA Section 210 meant that Osborne could sell the electricity produced by his turbines and the federal Production Tax Credit (PTC) paid Osborne $0.27 for each kilowatt hour his wind turbines produced. These policies were, however, insufficient to make these early wind farms economically viable, because the incentives only covered about half of the costs. After less than five years, lightning damage to the gear boxes forced Texas' first commercial wind farm to close.

In spite of the strong wind resources, the federal PTC, and access to the grid, these factors were not sufficient to promote wind development in Texas over the next two decades. Rather, the climate for wind development became favorable through shifts in the state-level electricity production and state policy advocates. Policy was one key driver. When the Texas Legislature passed Senate Bill 7 in 1999, signed by then Governor George W. Bush, they simultaneously deregulated the electricity industry and established Texas' first Renewable Portfolio Standard (RPS). This first RPS targeted 2,000 MW of installed renewable capacity by 2009.

Given the area's excellent wind resources, much of the initial wind was sited in windy yet remote West Texas. As wind power generation ramped up, it became clear that the existing transmission infrastructure was insufficient. Transmission congestion became a critical bottleneck limiting future wind development. Unlike other areas of the United States, the Texas electric grid is isolated from other states; there are no synchronous connections between ERCOT and other regional grids. This islanded system means that FERC does not have jurisdiction over the territory because there are no interstate connections. The boundaries and isolation of the Texas electricity grid made the need for rapid transmission development and more sophisticated grid controls essential, especially to accommodate periods of high wind. Early on in the Texas wind boom the lack of transmission-line access forced wind developers to select sites with less favorable wind resources, but with available transmission capacity. In Texas, the users of the electricity – or "load," in utility speak – to be transmitted through new lines pay for transmission system upgrades, but in the past the company developing the generation resource (such as a large-scale coal plant) would have sufficient collateral to secure financing for the construction of the new transmission lines. Many of the smaller wind companies did not have the financial resources to finance hundreds of miles of new transmission lines to connect remote wind sites to population centers, so transmission congestion rapidly emerged as a major bottleneck.

6.4.2 Turbines to Transmission: Texas Competitive Renewable Energy Zones

In an attempt to reduce this bottleneck to wind development, the Texas Legislature passed Senate Bill 20 in 2005, which required the PUCT to establish Competitive Renewable Energy Zones (CREZ) to help coordinate transmission planning and wind power development (Lasher 2008). The goal of this legislation was to promote wind development by solving the "chicken or egg problem," by first identifying the areas with the most promising wind resources and then supporting the planning and construction of new transmission projects to service those areas. ERCOT worked with the company AWS Truewind to identify twenty-five wind zones and then planned the necessary transmission lines to link the wind resources to the ERCOT grid and load centers like Dallas. ERCOT modeled the effects of the new wind zones and transmission lines to estimate the effects of these additions on system reliability, operations, and costs.

This study was then used by the PUCT (Docket 33672) to justify the creation of new zones to focus transmission development. The PUCT heard presentations from transmission developers on their plans for new lines and from wind developers on their wind power investments in the state. The PUCT was required to assess wind developers' financial commitment in wind energy in the PUCT's CREZ designations. Of the original twenty-five CREZ, nine were eliminated as there was no existing financial commitment from wind developers, and another eight were removed from consideration due to limited interest from wind developers (Public Utility Commission of Texas 2008). In October 2007, the PUCT established five CREZ through an interim order and ERCOT was tasked with conducting another scenario-based study to examine transmission plan optimization at four different levels of wind power development (Public Utility Commission of Texas 2008). While the interim order specified that all wind had to be sold within the ERCOT system, two of the new CREZ (CREZ Panhandle A and CREZ Panhandle B) were not located within ERCOT, but in the small region of Texas serviced by the neighboring Southwest Power Pool (SPP). Tensions arose over the question of whether the wind in CREZ Panhandle A and Panhandle B should connect to SPP or ERCOT (Lasher 2008). Ultimately this wind resource was connected to ERCOT.

When the PUCT order was approved in 2008, the PUCT identified more than ten high-priority transmission lines for areas already suffering from severe congestion. For example, in Scenario 2 studied by ERCOT, total costs of transmission were estimated at roughly $5 billion dollars for 2,335 miles of new 345 kV lines, or $426,000 per MW of capacity. The PUC considered the more aggressive plans with higher wind penetrations, but found them to be too risky and speculative given current technology and knowledge about grid integration.

Many of the early wind operators argued that they should gain priority access to the new transmission-line capacity. They argued that additional future wind generators should be curtailed first when transmission constraints occurred. The PUCT, however, elected not to adopt a dispatch priority rule. While FERC has approved the "anchor-tenant" concept, where initial projects finance some of the costs of transmission expansion to secure transmission rights, this has not been adopted in Texas. In Texas, transmission is paid for by users of the electricity (not the electricity generators), and regulators believed that granting priority access would leave later projects with less advantageous interconnection rights. Substantial investment in expanding the transmission network allowed for rapid wind development, although planning decisions on line size and location were often contentious, because everyone was aware of how each decision determined limits on the scale of future wind deployment. Later PUCT dockets addressed the selection of transmission providers (No. 35665), priority, and subsequent transmission-line sequencing (36801 and 36802). Each of these orders linked the PUCT, ERCOT, and transmission service providers in the development of the CREZ.

To build the transmission lines, the companies which were bidding to build the new transmission lines – the transmission service providers (TSPs) – studied the area, defined diverse routes for specific lines, identified owners of land that would need to be consulted, and held public consultations and informational meetings. The companies filed a Certificate of Convenience and Necessity (CCN) with the PUCT to ensure cost recovery. Assuming the information was acceptable, the PUCT would then approve the line and the level of investment that could be recovered through increases in electricity prices, or "cost recovery."

For transmission-line planning and siting, the TSPs would gather data from multiple constituents, including the counties, municipalities, landowners, Texas Department of Fish and Wildlife, the Texas Historical Commission, and other relevant parties. Data would include, for example, environmental information, irrigation pivot information, airport locations, communication towers, park and recreational areas, and information on historical sites. From this information, the TSP would propose several potential routes, identify and notify landowners along the different transmission-line alternative routes through mailings and newspaper announcements, and hold a series of public meetings in the project areas.

These data and a series of meetings form the basis for the CCN application to the PUCT. Once the project is approved by the PUCT, the TSP begins discussions with landowners on acquiring a right of way, crews survey the properties, and detailed engineering studies are performed. Finally, construction crews build towers and string power lines, and the line is put into service. The entire process averages five years (Cross Texas Transmission 2009).

While some complain that the multiple high-voltage 345 kV lines have led to "transmission fatigue," or communities becoming exhausted by the multiple ongoing efforts to site transmission lines, the additional transmission capacity built to support wind power development has also supported the development of shale gas resources in central Texas. The directional drilling and hydraulic fracturing technologies needed to develop shale gas require an energy infrastructure which did not exist before the CREZ development. While many supporters of wind power might not have chosen to promote development of additional hydrocarbons, Texas' focus on energy resource development remains a strong force among multiple stakeholders.

The CREZ transmission lines have helped wind power to serve electricity demand in populated areas of the state and have inadvertently and subsequently supported the development of the shale gas industry. The addition of significant wind power into the ERCOT electricity system has also forced grid operators to change how they manage the system to ensure reliability.

6.4.3 Integrating Wind into ERCOT

The scaling up of wind power in Texas has had a strong impact on regional grid operations and management. ERCOT operates a deregulated wholesale electricity market, so wind impacts electricity prices at different locations or "nodes," as well as other generators on the system (Public Utility Commission of Texas 2008). The variability of wind and its predictability both impact grid operations. When ERCOT commissioned GE and AWS Truewind to examine the effects of wind on power reliability needs with different levels of wind and load, their report modeled multiple scenarios: a 5,000 MW scenario, two 10,000 MW scenarios with wind resources sited at different parts of the state, and a 15,000 MW scenario. These different levels of system penetration were designed to help ERCOT better understand the operating requirements of different levels of wind development. The study found that high wind power penetrations would increase the need for flexibility and responsiveness in other generation sources. For example, to accommodate the increased wind power, the system would need to increase the capacity of fossil fuel power plants to ramp up more quickly; this could affect costs of grid operation and maintenance. The PUCT found this to be acceptable and focused on the report's claim that having wind on the system would reduce the overall spot price of electricity in electricity markets (Public Utilities Commission of Texas 2008). Other studies examined how wind would change the management of the grid to ensure reliability and evaluated how new wind forecasting methods could facilitate grid operations.

The PUCT also evaluated integration and reliability, and stated their belief in ERCOT's ability to integrate 18,000 MW of wind on the system by 2017. This represents 23 percent of projected peak system load. The PUCT also questioned how higher levels of wind integration would impact system reliability. In its decision to support the transmission for this high level of wind, the PUCT focused on system reliability but also cited positive impacts on strained water resources and reductions in criteria air pollutants (like oxides of sulfur or nitrogen) that would be associated with incorporating more wind into the electricity grid.

Along the way, Texas struggled with some high-profile challenges of integrating wind into the grid. A confluence of events on the afternoon of February 26, 2008 exposed the need to alter market scheduling and system operations to accommodate large levels of wind power generation. That day, several conventional generators were offline when the wind on the system dropped and load increased faster than expected in the afternoon. This caused ERCOT to call up its interruptible industrial and commercial clients – Loads acting as Resource (LaaR) – and curtail their demand. Later analysis of the event by the National Renewable Energy Laboratory (NREL) highlighted that it could have been avoided with better wind generation information, more accurate demand forecasts, and better scheduling of conventional units. While this event was resolved in less than two hours, it garnered a great deal of media attention – even though similar system incidents which did not involve wind received scant attention (Ela and Kirby 2008). This situation highlighted the need for integrating sophisticated weather predictions into ERCOT's system management; this integration is considered by some to be a key part of smart grid. Additional refinements to grid management have helped to avoid this type of event, even as installed wind power has grown.

When wind power is on the system, it displaces other generators. While most of the analyses in Texas highlight that additional wind on the grid displaces combined cycle gas turbines, periods of high wind and low load could affect coal or nuclear generators too. The PUCT specifically highlighted these considerations with regard to nuclear generators, which cannot easily be ramped up and down (Public Utility Commission of Texas 2009).

While policies to promote wind and ensure adequate transmission have been crucial, the larger economic context has been equally important in developing wind power for the Texas electricity industry. The marginal prices in the deregulated Texas electricity market are set by electricity produced from natural gas. As natural gas prices more than doubled from 2000 to 2009, the marginal cost of electricity in the ERCOT system also increased. With its policy support and transmission access, wind power became an economically viable – and profitable – resource. By 2007, more than 4,000 MW of wind power had been installed in Texas, two times greater than the original RPS goal for 2009. The Texas legislature revised the state's RPS upward, setting a target of 5,880 MW by 2015 and 10,000 MW by 2025. Both of these goals were surpassed by 2013. While installed capacity surged and lurched with the expiration and extension of the PTC in 2000, 2002, and 2004, it then increased steadily until 2013.

6.5 Upper Midwest: Wind and Transmission

Unlike Texas, where electricity transmission can be negotiated primarily among actors within the state boundaries, in the Upper Midwest the transmission grid connects with multiple states. Wind resources are linked by long-distance transmission lines to electricity demand centers in neighboring, or more distant, states. This makes large-scale wind development dependent on new institutional arrangements to facilitate interstate transmission planning and cost allocation. In this case study, we examine the shifting policy and institutional environments shaping wind and grid development in the Upper Midwest, with a focus on the Midcontinent Independent System Operator (MISO) region, which includes fifteen states in the Midwest and South.

In this case study we focus on the eleven Upper Midwest states which have been involved in coordinated transmission planning and wind integration since the mid-2000's. The Upper Midwest has some of the best wind resources in the world and long-distance transmission is crucial for developing the resource. With 13,000 MW installed, the Upper Midwest is similar to Texas in the amount of wind on the system (MISO 2010). All but one state in MISO's Upper Midwest region have adopted a RPS or a Renewable Goal, and together, meeting these policy goals will require upwards of 25,000 MW of wind on the system. While there has been quite a bit of research on state efforts to promote renewable technology through the state RPS or state renewable goals, deeper exploration of steps toward implementation such as grid planning and wind integration remains minimal (Bird et al. 2005; Rabe 2004; Rabe 2006; Rabe 2008). This case contributes to this area of research.

While state legislatures have adopted policies to promote renewables, implementation of wind and the required transmission expansion has required a new level of regional coordination and cooperation. Energy federalism in the United States means that state agencies have jurisdiction over critical energy policy decisions but there are no formal mechanisms to cooperate with neighboring states or integrate larger system interests into their decision criteria. Ratemaking and transmission system planning are two areas where the state-level PUCs have decision-making authority. PUCs use an established set of decision criteria to evaluate any system change, but some state PUCs are forbidden from considering benefits that may accrue outside of their state boundaries. The PUCs' criteria generally include the project's impacts on system reliability and benefit-cost analysis to evaluate the impact of the changes on rate payers within the PUC's particular state. These strict decision criteria can impede systemwide, intrastate decision-making and make transmission planning particularly challenging.

6.5.1 Building Wind Power in the Upper Midwest

Like Texas, the Upper Midwest has strong wind resources and has been home to windmills for more than 150 years. They were used from the 1870's to the 1930's by farmers and ranchers to pump water for homes, crops, and livestock, and for some farms. Wind was also used to produce electricity until the wind pumps were made redundant by the government-sponsored New Deal rural electrification programs in the 1930's–50's.

Unlike Texas, which could unilaterally adopt to promote wind power, developing wind power in the Upper Midwest is more politically complex. Individual state policies to promote wind development began in the 1980's, when Iowa passed a 1983 law requiring investor-owned utilities to buy 105 MW of wind power. The largest wind installation in the mid-1990's after the California crash was in Minnesota, where the state legislature brokered a deal with Northern States Power (NSP, now Xcel Energy). This deal allowed the utility to store nuclear waste in dry casks near one of its nuclear power plants as long as Xcel installed an additional 425 MW of wind power, with another 400 MW required by 2012. The rise of state RPS policies after 2000 and the on-again, off-again federal Production Tax Credit (PTC) also helped to promote wind development across the Upper Midwest. In the mid-2000's "Green–Blue" political coalitions emerged, linking the environmental movement and labor and passed RPS in many Upper Midwest states with the hope of simultaneously promoting rural and blue-collar economic development. In Iowa and Minnesota, wind development for rural economic development linked the environmental community and rural lawmakers, who are often at odds with one another on other issues. In North Dakota, the wind lobby had to compete with the lignite coal lobby, and the resulting Renewable Portfolio goal was weaker than in other Midwest states, as it is not binding. These initiatives initially enjoyed strong bipartisan support as Republican lawmakers like Minnesota's Governor Tim Pawlenty played a key role in supporting new RPS legislation. Although the situation changed after 2009, bipartisan support was crucial for passing this legislation.

Electricity markets and planning in the Upper Midwest are coordinated by the FERC-authorized Midcontinent Independent System Operator (MISO). By 2013, all eleven MISO states in the Upper Midwest, with the exception of Kentucky, had passed a RPS or goal and installed more than 12,000 MW of wind power, with 25,000 MW committed under state RPS policies. Most of the RPS policies in MISO states allowed both instate and out-of-state generation to count toward the RPS requirements. The exception was Michigan, which required that renewables be generated instate or owned by a utility operating in Michigan. In Illinois, only investor-owned utilities were obligated to participate in the RPS. In addition to RPS, states have used other policies to promote wind development. Some of the other state-level policies include corporate income tax credits to allow for accelerated depreciation of wind-related assets, special grant programs, low-interest agricultural loan programs, property tax exemptions, sales tax incentives, easements, green pricing programs, net metering, and public benefit funds.

Additionally, the Midwest Renewable Energy Trading System allowed for states (except Michigan) to comply with their state's RPS by purchasing Renewable Energy

Credits (RECs). These RECs could be associated with renewable energy produced outside of the state boundaries. For example, a Minnesota-based utility could purchase RECs from wind produced in North Dakota or a utility in Wisconsin could purchase RECs from wind generated in Iowa. This facilitated regional resource development of strong wind resources, but transmission soon became a barrier.

As early as the mid-2000's, Midwest states began to realize that wind development was being impacted by lack of transmission. Like Texas, sites with good wind resources are often located far from centers of electricity demand. However, while in Texas the Legislature, the system operator (ERCOT), and the PUCT were able to coordinate transmission planning, financing, and construction in that state, the same activity in the Upper Midwest required a novel and unprecedented level of interstate coordination. Historically, transmission planning, approval, and siting is under state jurisdiction and building transmission lines is a notoriously difficult activity (Vajjhala and Fischbeck 2007; Klass and Wilson 2013). Linking distant wind resources to demand requires that states decide how much transmission is needed, where it will be sited, and, importantly, how costs will be allocated between users. These estimates are based on a set of power system projections which embed estimates of future demand, resource development, and financing mechanisms. In traditionally regulated states, utilities would do the analysis and then present the results to the state PUC for approval and authorization of rate recovery from customers. In this process, state projections and processes for transmission planning could be different or even contradictory to one another. The first interstate transmission planning in the MISO region, the MISO Transmission Expansion Planning (MTEP) process helped to coordinate transmission planning for economic and reliability purposes (MISO 2014).

This initial effort to integrate wind resources began with a small subset of Upper Midwest states. Politicians realized that although transmission planning was vital to fulfill many of their energy policies, they were not directly engaged in transmission development. In September 2008, the governors of Iowa, Minnesota, North Dakota, South Dakota, and Wisconsin signed the Upper Midwest Transmission Development Initiative (UMTDI), forming a coordinating group that was supported by staff from MISO and included the Organization of MISO states, which represents state PUC interests at MISO. The goals of the UMTDI were twofold: (1) to create a multistate plan to guide development of new transmission lines to support renewable energy development; (2) to develop a cost allocation methodology to share the costs of new transmission across the states. For states like North and South Dakota, with large wind resources and relatively low demand, additional transmission lines to export wind power were critical for wind development. For Wisconsin, which had an RPS but had few wind resources of its own, interstate transmission lines would help to fulfill a political goal rather than offer instate economic development.

With the overall goals of connecting wind resources to the grid while reducing grid congestion and enhancing reliability, the UMTDI established a ten-person executive team comprising a member from each state's PUC and one from the state governor's office, along with multiple working groups. The executive team was tasked with coordinating key stakeholders, including state regulators, transmission companies, electric utilities, and independent power producers (National Wind Coordinating Collaborative 2008).

Several critical questions – slightly different to those addressed in Texas – faced the UMTDI, including: which zones had the best wind resources? Where should transmission development be prioritized? How could regional economic development be promoted, and should the transmission lines be sized for future large-scale export to other regions like the neighboring RTOs? The UMTDI also studied how to ensure renewable resources would be developed and examined optimal grid designs to promote access across all states. While Illinois was not a member of the UMTDI consortium, it is such an important part of the Upper Midwest electric system that it was included in all of the transmission analyses (National Wind Coordinating Collaborative 2008).

As in Texas, UMTDI plans were accompanied by detailed transmission analysis, modeling power flows on the existing system and estimating how new transmission lines would alter flows, as well as how they would influence future system costs, reliability, and grid operations. The UMTDI worked with MISO to develop different future transmission scenarios to estimate the effects of alternate configurations on system operation and estimate the benefits and costs of the new lines.

The UMTDI scoped twelve different scenarios and integrated stakeholder comments into the final selection of two scenarios. The UMTDI explored the possibilities of adding 25 GW of wind in the five study states, with twenty different energy zones, and nine zones in Illinois. The analyses examined providing electricity to the study states as well as exporting 10 GW of wind power and different configurations for transmission-line and wind resource development. The UMTDI led to a larger MISO-coordinated effort to identify and integrate "Unique Purpose Projects," which were projects targeted at fulfilling state RPS or low-carbon goals. This required a shift among MISO members, from supporting lines to enhance reliability to embracing larger policy goals. Coming to agreement on the future transmission scenario involved coordinating multiple different stakeholder viewpoints, termed "turbulence" in the May 28, 2010 FERC filing.

Issues of cost allocation, uncertainties of benefit-cost calculations, and a group of transmission owners who began a parallel analysis without informing other stakeholders affected the study process. The Regional Generation Outlet Study II (RGOS II) led to the region-wide transmission planning effort by the Midwest Governors' Association and MISO which mapped, planned, and eventually approved the seventeen high-voltage lines across the system (Figure 6.5). The resulting seventeen Multi-Value Project (MVP) lines, valued at $5.2 billion dollars with costs to be shared across the entire region, was the first multistate planning effort of its kind, approved in December 2011. For a line to be considered an MVP line it had to cost at least $20 million, be at least 100 kV, and help to meet reliability or economic goals or help MISO members to meet their Renewable Portfolio policy goals.

While the effort resembled Texas' CREZ process in many aspects, and MISO members learned from the CREZ process, the level of interstate coordination and negotiation across the multiple stakeholders in MISO states required an unprecedented level of regional cooperation. MISO staff calculated the costs and benefits of the new transmission lines for system reliability and efficiency. They also estimated the degree to which new transmission would help in meeting policy goals and estimated that the new lines would save the region $297–423 million each year through inexpensive western wind power from the Dakotas, which would displace more costly fossil sources.

MISO submitted its MVP plan for cost allocation across the region to the FERC and the FERC approved most parts of the MISO plan. (FERC 2010). However, several states and utilities were unhappy with the plan and took the FERC to court. The structure of state RPS made some parties more likely to sue. For example, in Illinois, rural electric cooperatives are exempt from state RPS requirements, and did not want to pay for the lines which they felt would not benefit them. Michigan argued that it would not be able to proportionally benefit from MVP lines due to the fact that it uses very little power from the MISO grid, as well as by virtue of its RPS which does not allow for interstate trading and requires Michigan utilities to count only instate renewable generation to meet the standard.

In June 2013, the Seventh Circuit Court of Appeals issued its opinion, upholding MISO's MVP process (2013). It found that the Illinois utilities would benefit from increased reliability and system savings and dismissed the claim that benefits and costs could have been calculated more accurately, recognizing that significant uncertainty exists. It also rejected Michigan's claims. This decision allowed for the MVP projects to continue: five years after the first UMTDI working group, MISO was able to move forward with the initial MVP lines.

6.5.2 Smarter Grids Across the Midwest for Wind Integration

While the MVP process to expand high-voltage transmission was underway, MISO was also working to better integrate wind into the grid operations and electricity markets. In the MISO region, wind power production is highest during winter nights, precisely when demand is lowest. As wind became a more important part of the generation mix, this mismatch began to cause problems. Because wind resources were the least expensive on the system – the wind "fuel" for turbine power production is free and wind power producers receive the federal PTC – wind power fed into the system whenever the turbines were spinning, and it was considered "self-scheduled" (MISO 2011). In some areas, high wind production created negative locational marginal prices, forcing other generators to cease production or wind to be curtailed. In 2010, the variability of wind resources and transmission capacity constraints forced more and more wind resources to be curtailed, with 4.2 percent of all wind-generated electricity curtailed across MISO in 2010 (Wiser and Bolinger 2013).[2] When wind resources needed to be curtailed, someone from the utility control room would call the wind operator and tell them to "dispatch down," with the system operator calling back sometimes hours later and allowing the wind plant back onto the system. When wind curtailment occurred, wind operators and contracting utilities lost money.

Integrating wind into electricity markets also posed challenges. Traditional generators like coal or natural gas plants bid into day-ahead electricity markets, specifying how much energy they can provide and at what price. These marginal cost curves are used by MISO to estimate market clearing prices and then run security-constrained economic dispatch models which schedule levels of electricity generation for the next day. When a scheduled plant does not meet the scheduled expectations, it is penalized. However, the variability of wind resources makes day-ahead predictions inaccurate. While wind forecasting has improved tremendously – MISO uses weighted sums of three independent weather models to predict wind – and wind resources are persistent in the short term, day-ahead predictions of wind are not accurate enough for day-ahead markets. Thus, wind bidding into traditional day-ahead energy markets would likely be penalized for not meeting its generation targets.

To better integrate variable wind resources into the electric grid, MISO developed the Dispatchable Intermittent Resources (DIR) program. The DIR program combined sophisticated weather models and turbine control systems to allow wind to participate in electricity markets and projects to be automatically dispatched up or down, depending on the system need. Before MISO developed the DIR, wind was not allowed to participate in electricity markets, and the market change required FERC approval. In creating the DIR, MISO used an extensive review process that included stakeholder engagement and multiple subcommittees to define a new class of generators and change the transmission tariff language.

When MISO filed its DIR tariff language to FERC in November 2010, stakeholders also filed their concerns. Unlike the contentious battles over the MVP transmission lines above, very few participants filed comments against the DIR. Rather, most of the filed comments were from groups – many from the wind industry – which supported MISO's DIR program changes. While FERC asked MISO for some minor DIR program clarifications, the changes were approved (MISO 2011). Unlike the contentious and costly transmission planning, wind integration into the markets was largely procedural.

---

[2] With the exception of the Northern States Power (Xcel) service territory, where only 1.2 percent of wind was curtailed in 2010.

Today, almost 80 percent of MISO wind resources are part of the DIR program and integrated into energy markets; the amount of wind power curtailed has decreased by half since 2011. Wind plants bid into day-ahead electricity markets and wind's bid level is "trued up" ten minutes before dispatch. This allows wind generators to fine-tune their bids to account for the actual wind conditions. Thus, the short-term persistence of wind projections is used to more accurately integrate wind into the energy market. In MISO, a smart grid for wind has included new integration of controls, technology, transmission, and data. While wind is still not allowed to participate in ancillary service markets which ensure power quality and reliability, ongoing research into new control systems could change this in the future too (Ela et al. 2014).

6.6 Germany: Wind and Transmission

Large-scale wind energy in Germany is a piece of a larger story of an ongoing societal transformation, from initial technology development to the society-wide *Energiewende*, or Energy Transformation. As in Texas and the Midwest, German policies began by incentivizing the creation and deployment of wind turbines, but this effort has recently expanded beyond just promoting renewables to a much larger national commitment to shifting the entire German energy system to a renewables-based system – with a goal of 80 percent of electricity to be generated from renewable power by 2050. Large-scale wind and a smarter grid play a central role in this transition.

Compared to Texas, the land area of Germany is small; Germany consists of less than 138,000 square miles, which is roughly half the size of Texas. The building and integration of wind power in Germany has been different from that seen in either Texas or the Midwest. Germany is a densely populated country which imports more than 70 percent of its energy resources. Germany's best wind resources are located in the north of the country and offshore, but most of the energy demand is in the industrial south. Given the distance between the northern wind resources and the southern demand, development of new North–South transmission lines has emerged as a critical national issue.

By the end of 2013, Germany had installed over 33,000 MW of wind power, meeting about 10 percent of the country's electricity demand; this amount is about half of the wind power installed in the entire United States, yet it is installed on less than 5 percent of the land area of the continental United States (European Wind Energy Association 2014). In addition to helping Germany change its energy system to respond to climate change, energy security and rural economic development are also important rationales for developing wind power.

German wind development is a key part of the national *Energiewende* that was adopted as official policy in 2011 by Angela Merkel's government. Rooted in a strong

German antinuclear movement that has been a powerful political presence since the 1970's, the *Energiewende* moves Germany to a renewables-based energy system without either nuclear or fossil fuels. When the German Green Party rose to prominence in the 1990's on a platform promoting – among other things – the end of nuclear power generation, some considered it an extreme position. As the party became part of the governing coalition, what was once a fringe idea gained mainstream acceptance. In 2002, the Schroder government made a decision to phase out all nuclear plants, although at this time no deadline was set. After the nuclear disaster in the Fukushima Daiichi plant in 2011, antinuclear sentiment soared and Angela Merkel's government announced the closing of all German nuclear plants by 2022; the *Energiewende* was passed and became official policy, promising an aggressive transformation of the energy system and targeting greenhouse gas reductions of 40 percent between 1990 and 2020.

6.6.1 Building Wind Power in Germany

While the current situation highlights the perils of rapid development and the additional complexity of a nuclear phaseout, the German government had been a leader in wind turbine research and development for decades (discussed in Section 6.2). In the 1980's Germany set goals for building 100 MW of wind, but installed capacity remained low. In the fall of 1990, the Bundestag (the German Parliament) spurred widespread deployment when they adopted the Electricity Feed-in Law (*Stromein-speisungsgesetz*). This feed-in tariff required that utilities connect wind projects to the grid and compensate the wind generators at 90 percent of the average electricity retail sale price. Coupled with low-interest twenty-year loans, these generous incentives spurred the first large-scale wind development in the country. Some Lander (the sub-national states in Germany) provided additional incentives for wind development. By 1991, Germany had installed 50 MW of wind power; by the end of 2000, this had grown to 1,750 MW. Initially, many of the wind projects were small; some were only two to three 500 kW turbines and many were owned by local cooperatives or local farmers. As project size grew, the turbine size increased to 1.1 MW by 2000; by this point many of the companies implementing these projects were financed by investors from outside the community where the turbine was located. The relative impact of the incentive package also grew over the decade; the feed-in tariff remained strong, but costs of wind turbines decreased by 40 percent. This evolution of project economics and finance made wind power profitable for investors, so further development occurred even in areas with less favorable wind resources.

In the late 1990's efforts to restructure the German electricity market further shifted the economics and financing of wind power. In February 2000 the Bundestag passed the Renewable Energy Law (*Erneuerbare-Energien-Gesetz* or EEG), which continued the feed-in tariff despite opposition from large utility companies. The effects of this law were to ensure investor confidence in wind and to encourage wind development in less windy areas by paying a higher feed-in tariff for development in these regions for the first five years and then gradually reducing it. For example, in 2002, more than 3,200 MW of wind power was installed in Germany. In 2004 the EEG was revised with lower prices for wind-generated electricity and a faster reduction in the feed-in tariff. Turbine size continued to grow too; the average size of turbines in 2002 was 1.4 MW, and by 2007 the average size was 1.9 MW.

Offshore wind development in Germany's North and Baltic Sea has been slower than onshore. Although the German government plans for an additional 25,000 MW of wind offshore and provides generous subsidies for offshore development, offshore wind power is costly to build and service, and until recently offshore wind development lacked a subsea transmission cable: transmission owners are responsible for building subsea cables, and this slowed development. Developing offshore wind requires a 400kV high-voltage network and the planned turbines are the largest yet, with 3–5MW turbines planned for the offshore wind fields. In 2014, 500 MW of offshore wind power was planned to be connected to the grid, with another 2,400 MW of wind power under construction (Offshore-Windenergie.net 2014).

Calculating the benefits – and costs – of new wind turbines remains contentious. For example, in 2006, it was estimated that German consumers pay an additional 3.3 billion euros, but that the value of electricity (5 billion euros), carbon savings and avoided air pollution (3.4 billion euros), and displaced fossil fuels (1 billion euros) results in a strong benefit-to-cost ratio. However, other analyses present just the costs – estimated at 23 billion euros in 2013 – without associated calculations of the benefits (Neubacher 2013).

6.6.2 Transmission to Connect Northern Wind to Southern Energy Demand

As in the United States, energy federalism is shaping the development of wind energy in Germany. The Federal Network Agency (*Bundesnetzagentur* or BNetzA) regulates Germany's electricity and gas markets and is responsible for coordinating transmission planning and network development plans to meet federal requirements. BNetzA is also responsible for assessing the environmental impacts of planned transmission projects. When Germany passed the *Energiewende* in 2011, it also passed the Grid Expansion Acceleration Act for Transmission Networks (NABEG) to try to help speed transmission planning and development.

While the federal government has been involved in making new transmission plans to connect northern wind resources to southern demand, the subnational German Lander and municipalities have important planning rights too. The German constitution guarantees Lander, municipalities, and municipal associations the rights to regulate local affairs, and transmission planning decisions have traditionally been made at the local level. As a result, in spite of strong central government support, the strong federalist nature of the German political system has made it exceedingly difficult to build the required North–South transmission lines.

Finally, in 2013, the government passed the Federal Requirement Plan for Transmission Networks (*Bundesbedarfsplan Übertragungsnetze*) to support the development of thirty-six new transmission lines and gave the BNetzA expanded authority to plan, site, and approve the lines. While actual development will be done with private transmission owners, expanded authority could help to facilitate and speed line development. However, this remains controversial. For example, the town of Meerbusch is planning to file a constitutional complaint, alleging that its municipal planning rights are being violated by the new federal law (Lang and Mutschler 2014).

6.6.3 *Integrating Wind (and PV) in the* Energiewende

Integration of renewable resources into the German grid is mandated by law. The policy of February 2000, "An Act on Granting Priority to Renewable Energy Sources" (Renewable Energy Source Act), gives all renewable-based generation explicit priority access to the grid. This means that when wind or solar plants are producing electricity, they are dispatched first and other generators are displaced. In addition to the rapid rise of wind, solar PV has increased, with more than 36 GW installed by April 2014. When wind and solar were small, this did not greatly affect grid operations or utility profits, but as they have become a more significant portion of the system, other generators – gas, coal, and even nuclear plants – have needed to ramp dispatch up or down to accommodate the renewable resources.

Together, the rapid rise in renewables and the priority dispatch order have affected the profits of other generators. Fuel for wind and solar plants is free and when they operate they can make the overall system cheaper, but this also undermines the business models of traditional generators and utilities. These losses have become significant because the fixed costs of a power plant are spread over each hour the plant operates. A gas-fired power plant in Germany, for example, needs to operate for at least 4,200 hours per year to be economically viable, and many have been shuttered. So integrating renewables has threatened the viability of many conventional generators. The more frequent ramping up and down reduces the life of conventional power plants, increases wear and tear, reduces profits, and sends long-term economic signals to utilities not to invest in conventional fossil fuel generation. German utilities E.ON and Vattenfall Europe Transmission, who absorb 42 percent and 38 percent of the share of wind power, as well as many other German utilities have seen their credit ratings drop, raising their cost to borrow capital. For the first time since 1949, German utilities are projecting losses: RWE posted losses of $3.8 billion and Vattenfall of $2.3 billion in 2013 (Lacey 2013).

Additionally, the German transmission system is managed as part of the European Electric Grid, and the lack of north–south transmission lines and the "renewables-first" dispatch policy has created electric spillovers which affect neighbors in the European Electric Grid. As electricity flows take the path of least resistance, congestion within the German transmission lines has meant that in times of high wind and solar production, German renewables operation has affected grid operations (and electricity prices) in neighboring countries. There have been increasing adverse impacts on links with Germany's neighbors in Poland and the Czech Republic on days of high wind and solar generation, and these unanticipated and "unscheduled power flows" have affected neighboring electricity operations and markets. To rectify this situation, grid operators from both the Czech and the Polish systems are working with the German grid operators to address the unscheduled power flows and loops. Grid planners are optimistic that the new north–south lines will alleviate this situation (Boldis 2013; Lang and Mutschler 2013).

The *Energiewende* is an aggressive model of national-level energy transformation, including electricity use as well as other forms of energy. While its goals are ambitious – to some, even audacious – it has spurred wind and solar generation more quickly than previously imagined possible. Not surprisingly, implementation and integration challenges in the first few years have been significant. Outside forces, like the weak price for carbon credits in the European Emissions Trading System and the rise of shale gas in the United States, and resulting exports of cheaper North American coal, meant that cheaper coal-fired power generation in Germany increased by 8 percent in the first half of 2013 (Neubacher 2013). Ironically, increased wind and solar capacity, coupled with the phasing out of nuclear, has been accompanied by increased coal use and higher greenhouse gas emissions in Germany. So in the near-term some pieces of the larger energy transition have, at least temporarily, moved greenhouse gas emissions in the wrong direction. These additional tensions are having an impact on the development of wind power and smart grid in Germany.

6.7 Future of Wind and Smart Grid

In this chapter, we have explored the interactions and dependencies between development of wind power and smart grid in three different regions. In each of these areas large-scale wind power has been changing the electric system planning, operation, and politics and smart grid development has been intricately linked with wind integration. The locus of control for developing wind power has shifted from turbine developers to policymakers, electric utilities, grid operators, and around again. The growth of wind power has forced the energy system to evolve in new ways. It has changed the planning, financing, and operation of the transmission grid and simultaneously forced the development of a smarter grid. Integrating wind into the electric system requires operators to use detailed weather information and new control systems to manage resource variability.

While the past decade of large-scale wind development in the United States was spurred by state Renewal Portfolio Standards (RPSs), it was also shaped by outside factors and technology developments. In Texas, wind power initially enjoyed a comparative cost advantage to other generation resources. This cost advantage has been eroded by low natural gas prices due to shale gas development using hydraulic fracturing, which has caused natural gas prices to plummet from $9–13/thousand cubic feet to $3–5 per thousand cubic feet. In the MISO region, state RPS mandates and goals initially spurred wind development, but the strong wind resources have made wind power a cost-competitive resource. Thanks to new smart grid control systems in the MISO region, wind bids directly into day-ahead electricity markets and is automatically controlled when it is needed. In Germany, the *Energiewende* continues to drive both onshore and now offshore wind development, but parallel federal policies to facilitate transmission-line development are also crucial to alleviate wind integration challenges.

In each of these cases, the control of wind on the system has benefited different parties, from wind developers to energy consumers. Additionally, large-scale renewables integration has impacted electric system economics and shifted traditional electric system boundaries in unanticipated ways. Creating large-scale wind power has affected incumbent actors in unexpected ways, too. In Texas and the Upper Midwest, wind power has lowered the cost of energy to the entire system, but the additional transmission costs have not been cheap. In Germany, the recent losses posted by traditional utilities highlight just how much renewables like wind and solar have shifted the economics of the power grid. Utilities dramatically refer to this type of system loss as the "death spiral" (Lacey 2013), and predict they could lose billions of dollars per year with increased renewables penetration.

Large-scale renewables have also been part of unexpected system changes. A major struggle in Germany has emerged as coal consumption has increased in parallel with renewable generation, in large part due to the simultaneous end of nuclear that is also an aspect of the *Energiewende*. This has created a paradox of more carbon-free renewables creating more carbon from the electric system. This highlights the challenges in predicting multiple system shifts from the development of new technologies and smarter grids.

The three case studies we explored in this chapter demonstrate both the synergies and tensions of wind power and smart grid development. In the United States, in both Texas and the Upper Midwest, the combination of strong wind resources and favorable state policies have resulted in wind becoming a critical part of the electricity system, including its integration into electricity markets. Tensions include planning and paying for new transmission lines. For Germany, even though wind and smart grid development are embedded within the larger *Energiewende*, renewables development has challenged the foundations of energy system planning and system operation.

Enabling large-scale renewable energy is one of the most prevalent promises of smart grid, and together wind farms and a smarter grid are helping to create new electricity systems. The stories of the coevolution of wind power and smart grid highlight how growth in one technology can reshape an entire system. Wind and a smarter grid have also shifted energy markets and power system rules; reordered the institutional priorities of utilities, regulators, grid managers, and other established actors; shifted system politics; and created opportunities for new entrants in the electricity sector. Technology to harness the wind has played an important role in human history, from grinding grain in feudal food systems to pumping water for settlements and train transportation in nineteenth-century North America. Now wind power and smart grid are playing a critical role in a larger societal transition to a more sophisticated and environmentally friendly electricity system.

References

Bird, L., M. Bolinger, T. Gagliano, R. Wiser, M. Brown, and B. Parsons. (2005) Policies and Market Factors Driving Wind Power Development in the United States. *Energy Policy*, 33, 1397–1407.

Boldis, Z. (2013) Czech Electricity Grid Challenged by German Wind. *Europhysics News*, 44, 16–18.

Cross Texas Transmission. (2009) Transmission to Deliver Renewable Energy. LS Power. class4winds.org/seminar/Willick_CrossTexas.pdf

Ela, E., V. Gevorgian, P. Fleming, Y. C. Zhang, M. Singh, E. Muljadi, A. Scholbrook, J. Aho, A. Buckspan, L. Pao, V. Singhvi, A. Tuohy, P. Pourbeik, D. Brooks, and N. Bhatt. (2014) *Active Power Controls from Wind Power: Bridging the Gaps*. Golden, CO: NREL. NREL/TP-5D00-60574, www.nrel.gov/docs/fy14osti/60574.pdf

Ela, E. and B. Kirby. (2008) ERCOT Event on February 26, 2008: Lessons Learned. *National Renewable Energy Laboratory*, Golden, CO: National Renewable Energy Laboratory. NREL/TP-500-43373 www.nrel.gov/docs/fy08osti/43373.pdf

European Wind Energy Association. (2014) *Wind in Power: 2013 European Statistics*. Brussels, Belgium: EWEA.

FERC. (2005) Interconnection for Wind Energy. In *Docket No. RM05-4-001*. Washington, DC: FERC. www.ferc.gov/EventCalendar/Files/20051212171744-RM05-4-001.pdf

FERC. (2010) Order Conditionally Accepting Tariff Revisions. In *ER10-1791-000*, Washington, DC: FERC. www.ferc.gov/whats-new/comm-meet/2010/121610/E-1.pdf

FERC. (2011) Transmission Planning and Cost Allocation by Transmission Owning and Operating Public Utilities. In *18 CFR Part 35*, 18 CFR Part 35. Washington, DC: FERC. www.ferc.gov/whats-new/comm-meet/2011/072111/E-6.pdf

Galbraith, K. and A. Price. (2013) *The Great Texas Wind Rush*. Austin, TX: University of Texas Press.

Heymann, M. (1998) Signs of Hubris: The Shaping of Wind Technology Styles in Germany, Denmark, and the United States, 1940–1990. *Technology and Culture*, 39, 641–670.

Hirsh, R. F. (1999) *Power Loss: The Origins of Deregulation and Restructuring in the American Electric Utility System*. Cambridge, MA: MIT Press.

Illinois Commerce Commission et al. v FERC. (2013) http://media.ca7.uscourts.gov/cgi-bin/rssExec.pl?Submit=Display&Path=Y2013/D06-07/C:11-3421:J:Posner:aut:T:fnOp:N:1148803:S:0

Kaldellis, J. K. and D. Zafirakis. (2011) The wind energy (r) evolution: A short review of a long history. *Renewable Energy*, 36, 1887–1901.

Klass, A. B. and E. J. Wilson. (2012) Interstate Transmission Challenges for Renewable Energy: A Federalism Mismatch. *Vanderbilt Law Review*, 65 (1801–1873). www.vanderbiltlawreview.org/2012/11/interstate-transmission-challenges-for-renewable-energy-a-federalism-mismatch/

Lacey, S. (2013) *This is What the Utility Death Spiral Looks Like*. San Francisco, CA: Green Tech Media. http://www.greentechmedia.com/articles/read/this-is-what-the-utility-death-spiral-looks-like

Lang, M. and U. Mutschler. (2013) *Bundesrat Passes Federal Requirement Plan for Transmission Networks*. Berlin: German Energy Blog. www.germanenergyblog.de/?p=13250

Lang, M. and U. Mutschler. (2014) Meerbusch Announces to File Constitutional Complaint Against Law Accelerating Expansion of Electricity Grids. Berlin: German Energy Blog. www.germanenergyblog.de/?p=13830#more-13830

Lasher, W. P. (2008) The Development of Competitive Renewable Energy Zones in Texas. Paper presented at the *Transmission and Distribution Conference and Exposition, 2008*. Chicago, IL, April 21–24, 2008. ieeexplore.ieee.org/xpls/abs_all.jsp?arnumber=4517254

Meegahapola, L. and D. Flynn. (2010) Impact on Transient and Frequency Stability for a Power System at Very High Wind Penetration. Paper presented at the *Power and Energy Society General Meeting, 2010 IEEE*, July 25–29, 2010, Minneapolis, MN. ieeexplore. ieee.org/xpl/articleDetails.jsp?tp=&arnumber=5589908&queryText%3DImpact+on+Transient+and+Frequency+Stability+for+a+Power+System+at+Very+High+Wind+Penetration

MISO. (2010) *Regional Generation Outlet Study*. MISO, Carmel, Indiana. www.misoenergy.org/Library/Repository/Study/RGOS/Regional%20Generation%20Outlet%20Study.pdf

MISO. (2011) Compliance Filing of the Midwest Independent Transmission System Operator, Inc. regarding Dispatchable Intermittent Resources Docket No. ER11-1991-00. FERC: FERC.

MISO. (2014) MISO Transmission Expansion Planning (MTEP). MISO, Carmel, Indiana. www.misoenergy.org/Planning/TransmissionExpansionPlanning/Pages/TransmissionExpansionPlanning.aspx

Musgrove, P. (2010) *Wind Power*. Cambridge, UK: Cambridge University Press.

National Wind Coordinating Collaborative. (2008) Transmission Update. nationalwind.org/wp-content/uploads/2013/05/NWCCTransmissionUpdateDec08.pdf

Neubacher, A. (2013) Germany's Defective Green Energy Game Plan. *Der Speigel*. www.spiegel.de/international/germany/commentary-why-germany-is-waging-its-green-revolution-wrong-a-929693-druck.html

NREL. (2008) *20% Wind Energy by 2030*. Golden, CO: NREL. www.nrel.gov/docs/fy08osti/41869.pdf

NREL. (2014) WINDExchange. apps2.eere.energy.gov/wind/windexchange/wind_installed_capacity.asp

NREL & AWS Truepower. (2014) United States (48 Contiguous States) – Wind Resource Potential Cumulative Rated Capacity vs. Gross Capacity Factor (CF). Golden, CO: U.S. Department of Energy. apps2.eere.energy.gov/wind/windexchange/pdfs/wind_maps/us_contiguous_wind_potential_chart.pdf

Offshore-Windenergie.net. (2014) Wind Farms. www.offshore-windenergie.net/en/wind-farms

Porter, K., S. Fink, C. Mudd, and J. DeCesaro. (2009) *Generation Interconnection Policies and Wind Power: A Discussion of Issues, Problems, and Potential Solutions*. Golden, CO: NREL. http://www.nrel.gov/docs/fy09osti/44508.pdf

Price, T. J. (2013) Blyth, James (1839–1906). In *Oxford Dictionary of National Biography*. Oxford, UK: Oxford University Press. www.oxforddnb.com/view/printable/100957

Public Utility Commission of Texas. (2008) Docket No. 33672, Commission Staff's Petition for Designation of Competitive Renewable Energy Zones. Austin, TX: Public Utility Commission of Texas. www.ettexas.com/projects/docs/PUCTFinalOrderonCREZPlan_10-07-08.pdf

Public Utility Commission of Texas. (2009) Project No. 34577, Proceeding to Establish Policy Relating to Excess Development in Competitive Renewable Energy Zones, Order Adopting Amendments to §25.174. Austin, TX: Public Utility Commission of Texas. www.puc.texas.gov/agency/rulesnlaws/subrules/electric/25.174/34577adt.pdf

Rabe, B. G. (2004) *Statehouse and Greenhouse: The Evolving Politics of American Climate Change Policy*. Washington, DC: Brookings Institution Press.

Rabe, B. G. (2006) *Race to the Top: The Expanding Role of the U.S. State Renewable Portfolio Standards*. 38. Washington, DC: Pew Center on Global Climate Change. www.pewclimate.org/global-warming-in-depth/all_reports/race_to_the_top/index.cfm

Rabe, B. G. (2008) States on Steroids: The Intergovernmental Odyssey of American Climate Policy. *Review of Policy Research*, 25, 105–128.

Schwartz, M., D. Heimiller, S. Haymes, and W. Musial. (2010) *Assessment of Offshore Wind Energy Resources for the United States*. NREL. Golden, CO: U.S. Department of Energy. NREL/TP-500-45889, www.nrel.gov/docs/fy10osti/45889.pdf

Scientific American. (1890) Mr. Brush's Windmill Dynamo. *Scientific American*, 383, 389.

U.S. Department of Energy. (2011) *History of Wind Energy*. Washington, DC: U.S. Department of Energy. energy.gov/eere/wind/history-wind-energy

Vajjhala, S. P. and P. S. Fischbeck. (2007) Quantifying Siting Difficulty: A Case Study of US Transmission Line Siting. *Energy Policy*, 35, 650–671.

Vestergaard, J., L. Brandstrup, and R. Goddard. (2004) Industry Formation and State Intervention: The Case of the Wind Turbine Industry in Denmark and the United States. In *Academy of International Business Conference Proceedings*, Southeast USA Chapter. www.hha.dk/man/cmsdocs/publications/windmill_paper2.pdf

Wiser, R. and M. Bolinger. (2013) *2012 Wind Technologies Market Report*. Berkeley, CA: Lawrence Berkeley National Laboratory. emp.lbl.gov/sites/all/files/lbnl-6356e.pdf

Xu, Z., M. Gordon, M. Lind, and J. Ostergaard. 2009. Towards a Danish Power System with 50% Wind: Smart Grids Activities in Denmark. Paper presented at the *Power & Energy Society General Meeting, 2009*, July 26–30, Calgary. http://ieeexplore.ieee.org/xpls/abs_all.jsp?arnumber=5275558&tag=1

7

Community and Small-Scale Grid Innovation

7.1 The Promise of Local Control

One of the many promises of smart grid that we outlined in Chapter 2 is the potential for more local electricity generation and community control. Local distributed generation (DG) offers new possibilities for community engagement and ownership in electricity systems. This chapter explores the tensions and opportunities for smart grid to contribute to local and small-scale electricity system initiatives. In this chapter we explore how smart grid is shaping (and being shaped by) small-scale energy initiatives in which communities, individuals, and organizations engage in electricity systems planning at the local level. The chapter includes discussion of community-based electricity systems and microgrids. Microgrids are generally defined as miniature versions of the larger electricity system, and often microgrids have the potential to separate from the larger grid system – a capacity referred to as "islanding." Microgrids often, but not always, emerge alongside community-based initiatives led by *locavolts*, people who seek to build self-reliance through local control of their energy systems.

Local power and microgrid initiatives are emerging in many different contexts. Investments in these small-scale or community-oriented initiatives include efforts by municipalities trying to take control of their electricity systems, as well as internationally supported initiatives where governments are trying to encourage local control. These local electricity systems include initiatives on college campuses, military installations, and industrial facilities.

Small-scale and community initiatives, when examined as individual case studies, demonstrate many of the promises, pitfalls, and tensions associated with larger-scale notions of smart grid. In considering the development of small-scale grid initiatives, one of the dominant tensions relates to whether the grid should be designed to promote and support more centralized or more decentralized electricity systems. Many of the risks and benefits, promises and pitfalls of decentralized systems become clearly evident in small-scale community initiatives. A community-based electric grid, for example, may enable the community to keep the lights on when a weather-related disruption cuts off power to surrounding communities. At the same time, a community-based electric grid with dynamic pricing could expose electricity customers to economic challenges resulting from greater variability in electricity prices than they might experience as part of a larger-scale system, because electricity prices fluctuate with demand.

This simultaneous potential for both benefits and risks explains how opposition to innovative small-scale grid projects can sometimes be grounded in the traditional business model and public service mandates to minimize people's exposure to risk. But depending on which risks and benefits are prioritized (e.g. economic, political, social, cultural, or technological) and who is bearing the risks or reaping the benefits, these initiatives can be viewed favorably or negatively by different key actors.

Implementing smart grid at any scale involves a diverse portfolio of potential technologies, as discussed in Chapter 3. The technologies that are most central to any specific community or small-scale grid initiative, whether it be storage technology or responsive islanding technology that automatically separates from the main grid when power goes out, depend on the function, structure, and motivation of the specific project. For some actors (including individual locavolts who are eager for engaged independent involvement in their electricity generation) the promises of small-scale, community engaged projects may far exceed potential pitfalls, while for other actors (including conventional investor-owned utility companies whose business model is based on selling power they distribute), the pitfalls may exceed the promises.

Although each small-scale initiative is distinctive in how it relates to broader notions of creating a smarter grid, all small-scale or community initiatives share some important commonalities in terms of advantages and disadvantages, central technologies, and key actors whose interests are well served, or not. We begin this chapter by first defining several key terms, including community-based energy, microgrid, nanogrid, locavolts, and prosumer. We then highlight the promises and pitfalls most obvious in small-scale electricity system initiatives, identify the technologies that are most central in these smart grid approaches, and describe the key actors and their interests most directly impacted by community-based or small-scale projects. We then describe in detail three different small-scale community initiatives that illustrate the diversity of approaches and the tensions among various actors as they attempt to achieve the promises and avoid the pitfalls. While many industrial parks, college campuses, and military bases are also developing distributed generation, in the selection of these three cases we focus primarily on communities.

The first case describes the ongoing (at the time of writing) community struggle to municipalize the electricity system in Boulder, Colorado. Our second case explores the neighborhood-scale Pecan Street smart grid project in Austin, Texas, which is a partnership between the public and private sectors supported primarily by the federal government. Our third case explains attempts toward energy independence and increased system reliability in Bornholm, Denmark, supported primarily by the European Union. After telling each of these stories, we summarize the commonalities and differences across cases, and explore both the opportunities and challenges involved in integrating these smaller systems into the overall vision of smart grid. Finally, we discuss how each of these cases, as well as some other organizationally specific small-scale grid initiatives at universities, in the military, and in other organizations, demonstrate interest in gaining greater local control of energy systems.

7.1.1 Defining Key Terms

In this introductory section we define a few key terms and provide some examples to clarify what we mean when we use them: community-based energy, microgrid, nanogrid, locavolts, and prosumer.

Community-based energy. Multiple new and creative ways of structuring community-based energy systems are emerging. Enhanced community engagement is being formalized through multiple approaches to giving local users greater control of electric systems which operate within existing institutional frameworks and rely on coordination between the public and private sectors. These initiatives often encourage small-scale solar and wind energy development, and emphasize legal ownership of the value-added product. For example, Windustry, an NGO that encourages "renewable energy solutions," defines community wind simply as "a community-owned asset" (Windustry 2014). In 2014, its annual Community Wind Innovator Award went to Chris Diaz at Seminole Financial Services, who has developed innovative solutions for the financial challenges facing community wind throughout the United States (Goldman 2011; Seminole Financial Services 2014,).

Community-based grid innovation is occurring in communities throughout the world. German development of smart grid is not limited to the large-scale wind projects described in the previous chapter. Rather, community-based renewable energy systems in Germany continue to develop rapidly (NAW Staff 2013; NAW Staff 2014b; NAW Staff 2014c), with robust participation from well-known companies such as Siemens and Vestas. And the enthusiasm is not limited to commercial actors. Communities are eager for what they perceive as an opportunity to generate value that stays onsite as increased income for individual residents, increased profits for local companies, avoided fuel costs, and increased taxes for local municipalities and states (NAW Staff 2014a; Krause 2013). Despite the costs to utilities (see Chapter 4), the German experiment with community-based energy has caught the attention of Southeast Asian nations, generating, for example, a cooperative program intended to help Thailand build on German successes with community-based renewables (Ministry of Foreign Affairs Thailand 2013).

Community-based solar initiatives, such as the system of rooftop solar panels that now provides electric power for the municipal building, the public works building, and three public schools in Warren, New Jersey (Independent Press 2011; Seminole Financial Services 2014), are seeing some success in the United States. The California Energy Commission has embraced community-based solar to the extent of preparing and distributing a guide for local communities (California Energy Commission 2009). California, which is experiencing consumer demand for improved energy performance among home buyers, formatted the guide as a tool kit to help municipalities partner with builders and other private sector actors, as well as the state government, to develop community-based solar energy. In the Midwest, Xcel has proposed a community solar gardens program that would allow Minnesota customers to participate in the development of solar energy projects, even if they are unable to install solar PV on their own roofs (Sustainable Business 2013; Xcel Energy 2014). Xcel proposes giving customers who participate in the program a credit on monthly electricity bills, drawn from their portion of the energy generated in the solar gardens.

Although U.S. development lags, supporters claim that community-based wind delivers significant local benefits (NAW Staff 2014a; Windustry 2014). Minnesota's Community-based Energy Development (C-BED) legislation illustrates another approach that relies on preexisting jurisdictional frames and relationships with large electricity providers, and has formalized a set of community-based options for wind energy development (Minnesota Department of Commerce 2013). The legislation provides financial support, or tariffs to encourage wind development at a scale that individual communities can handle, and includes the following requirements (Minnesota Department of Commerce 2013):

- 51 percent of the revenues from the power purchase agreement must flow to Minnesota-based owners and other qualifying local entities.
- No single wind project investor can own more than 15 percent of a project consisting of two or more wind turbines, except for local governments which may be the sole owners of community-based projects.
- The project must have a resolution of support adopted by the county board of each county in which the project is to be located, or, in the case of a project located within the boundaries of a reservation, the tribal council for that reservation.
- All owners of property traversed by transmission lines serving the project must be given the opportunity to invest.

This program is intended to make access to the energy system accessible to communities that desire greater control, yet have neither the inclination nor the resources to join the locavolt movement. Of course, some observers claim that C-BED is nothing more than a smokescreen for the large-scale wind energy development discussed in the previous chapter. Thinking about the difference between a two-turbine wind energy system built for a distribution center in Tracy, California (Seminole Financial Services 2014) and some of the larger C-BED projects should give some indication of how difficult it is to pin down the boundaries of this category. For this chapter, our interest in community-based energy is limited to relatively small initiatives, ranging from those that could power a single community building to those that could power a small city, such as Boulder, Colorado (our first case in this chapter). Our interest is further focused on community-based initiatives that are closely interconnected with smart grid.

There are also many emerging non-formalized community initiatives that are more grassroots in nature. In Massachusetts several towns have independently decided to build wind turbines and/or community-based solar installations. Also, in some places groups of residents without south-facing roofs have pooled their resources to install solar panels on a local community location instead of on their individual roofs.

Microgrid. Because there are so many definitions of "microgrid," we will start by providing the definition we use. First, while every microgrid is small, not every microgrid is "smart," and we are focused on the smart ones. We follow Mariam et al. in defining microgrids as "single electrical power subsystems associated with a small number of distributed energy resources" (Mariam 2013b). This DG can be from both renewable and/or conventional sources, and includes PV, small-scale wind turbines, micro-hydro, internal combustion engines, natural gas turbines, and microturbines, which are managed in a coordinated way to create a "cluster of loads" (Mariam 2013b). In smart microgrids, DG is linked with power electronic interfaces that provide users and system operators with the "flexibility to operate as a single aggregated system maintaining the power quality and energy output" (Mariam 2013a). Smart microgrids can operate as a single load, which can appeal to utilities and other system operators. At the same time, they appeal to customers by enabling them to meet electrical requirements "locally, supply uninterruptable power, improve power quality, reduce feeder loss, and provide voltage support" (Mariam 2013a). Put simply, the microgrids in which we are interested are grids that are equipped to integrate a wide variety of DG and renewable energy sources and can also "island" and operate independently of the larger grid. The diesel-powered backup generator in a grocery store parking lot does not qualify. The U.S. Department of Defense (DoD) Smart Power Infrastructure Demonstration for Energy Reliability and Security (SPIDERS) system does (Perera 2012).

Nanogrid. The term "nanogrid" refers to an approach to the electricity system that takes the microgrid emphasis on DG even further, down to a single load or actor. Bruce Nordman described a nanogrid as having "at least one load . . . and at least one gateway to the outside," (Nordman 2010). He and colleagues list a nanogrid's most basic components as controller, load, and gateway (Nordman, Christensen, and Meier 2012). As with most aspects of smart grid, definitions are still fluid, and cover a wide range of options. Navigant Research, for example, defined nanogrids as "100 kW for grid-tied systems and 5 kW for remote systems" (Hardesty 2014; see also Asmus and Lawrence 2014). Lawrence Berkeley National Laboratory described a nanogrid as having "at least one load or sink of power, a gateway to the outside, and a controller to distribute power, using price signals to mediate supply and demand. It is the most effective way to integrate local renewable generation and storage, and it incorporates features such as peer-to-peer power exchange, bidirectional power flow, and managed distribution to loads" (Chen 2012b). For our purposes, we follow the relatively broad definition of nanogrids as small microgrids (Hardesty 2014; Lundin 2014b), and add on an institutional consideration: microgrids can combine multiple actors into a coordinated configuration, but a nanogrid has a single decision maker. A nanogrid could refer to a personal computer and the USB-powered device connected to it, or it could refer to one of the DoD's SPIDERS installations.

Locavolt movement. The Locavolt movement focuses on changing the locus of control of the energy system from large companies to individual *prosumers*. Although it also aims to produce a more resilient electric grid and improve environmental quality, its primary purpose is individual empowerment. Author, journalist and energy expert Peter Asmus is widely credited with coining this term to describe a movement of people who seek to "generate power right in their own homes and neighborhoods" (Asmus 2008a). The analogy with the more familiar "locavore" movement among foodies is intentional. In response to naysayers who worry about loss of power quality and reliability, Asmus provides examples that range, both spatially and politically, from California to Iowa. He uses these examples to demonstrate the diverse ways that locavolts "secure reliable supplies in times of emergency" at the same time as enhancing their own self-reliance on an everyday basis. Technologies associated with smart grid, such as telecommunication advances and conversion devices, have enabled locavolts to tap into renewable resources that are locally available. For example, community wind projects dominate the locavolt movement in rural Minnesota and Iowa, whereas the locavolt movement in California is turning to a combination of rooftop PV and small wind turbines, with plug-in hybrid cars providing storage (Asmus 2008b).

Locavolts are part of a movement that relies on individuals who are willing and able to invest significant time, money, and other resources into achieving their goal of locally controlled electric grids (Mayer-Schonberger 2006; Endres 2009). Along with many of the new social movements, locavolts are a technologically savvy group with anti-corporate inclinations (Juris 2005; Carvalho 2012). Their organized actions rely on interactive media that are produced and distributed using computers, the internet, and social media, whose data can be widely accessed by anyone interested (Webster 2001). These media have shifted opportunities for public engagement in system-level changes. For example, face-to-face meetings and paper-based campaigns have taken a back seat to chat rooms, blogs, and Twitter feeds. The central point here is that locavolts both generate and use power locally, and they use communication technologies connected through the internet to make it happen. This is not to claim that locavolts never participate in more traditional organizing activities. They combine use of these interactive media with more traditional on-the-ground organizing and activism. For example, the Local Clean Energy Alliance held a festival titled "Locavolts Unite" on November 13, 2008 in San Francisco (Local Clean Energy Alliance 2013). The organization has a large web-based membership, regularly conducts policy briefings, and uses the internet to coordinate campaigns supporting or opposing state legislation related to energy.

Prosumers. Our interest in the locavolt movement grows out of its potential to nurture the cadre of prosumers who are integral to smart microgrids. As we discussed in Chapter 2, prosumers are fully engaged in the energy grid, even to the extent of participating in basic innovation. Rather than divorcing themselves from the system, they are involved in changing it through producing electricity. Their empowerment grows out of their belief that they have the potential to build a better system, rather than from deep alienation. Although we recognize there are individuals for whom the electricity system, no more or less than any other system, represents a fundamental threat, this is not our focus. We are interested in the emergence of the prosumer because it represents a fundamental shift in how energy system actors identify themselves and characterize each other. It also indicates the importance of providing more participatory opportunities for those who demonstrate an interest in, and a sense of responsibility for, the future energy system. In this chapter, the prosumers we focus on identify themselves as locavolts, whose political activities are intended to build a more horizontally organized energy system.

7.2 Promises and Pitfalls

As a demonstration of the recurring tensions between centralization and decentralization, small-scale grid initiatives are especially likely to deliver certain benefits and also pose specific risks. Small and independent projects could promise their participants increased reliability if they could enable continued functionality when the larger system goes down. Superstorm Sandy provided an opportunity for microgrids in the Eastern United States to demonstrate their enhanced resilience. Princeton University's microgrid, for example, ensured that essential electricity services were restored almost immediately, while most of the region remained without power.

These small-scale initiatives also promise economic benefits, including enhanced energy security and a stronger local economy. Potential benefits range from cutting municipal costs by installing more efficient street lighting, to the promise of creating new jobs for local residents, to building a smart and resilient grid.

Small-scale community-based initiatives may be somewhat limited in their contribution to environmental health because of scale issues and the fact that many incorporate fossil generation like diesel generators or natural gas-fired microturbines. Still, proponents of small-scale electricity often argue that, by choosing to increase the proportion of electricity generated by renewable resources and by decarbonizing the electricity production for a single community, they have improved environmental quality by contributing to climate change mitigation, as well as reducing local emissions of air and water pollutants. Again, this depends on what sources of power generation are included in the microgrids and how they are operated.

One of the most significant promises of community-based small-scale local initiatives is the enhancement of citizen engagement. Advocates of community-based electricity note that decisions are made at the local level by the people whose lives are most directly impacted by those decisions. Community-based smart grid initiatives could offer high-quality public participation opportunities, and enable people to develop new connections with their energy system. Community power projects have the potential to empower electricity users by enabling them to become self-reliant prosumers; people who directly influence the system, rather than passive consumers. For some municipalities, local control could also bring independence, and with that comes power and freedom from the big utilities or from the larger systems.

Not everyone is persuaded that microgrids and community-based electricity systems are a positive development for society. Some actors point out the dangers posed if communities are allowed to go their own way and gain independence through locally controlled electricity systems. Incorporating more community-based electricity into the system could harm the larger system's existing reliability through loss of redundancy or underinvestment in communal transmission networks. Power quality is one of the most often cited issues of small-scale grid initiatives. Because distributed renewable resources are highly dependent on environmental factors, their variability has introduced some power quality problems into the system (Mariam 2013a).

Small-scale power projects also have potential to exacerbate societal inequalities, as richer communities are able to opt out of the communal electricity grid but poorer communities do not have the resources to do so. Further, individual communities may be financially overburdened if risks cannot be shared across the larger system. As noted especially in Chapters 3 and 4, smart grid requires huge investment in new technology – investment that may overextend local communities.

And despite all the hype about locally sourced renewable energy, local power could actually harm environmental quality. If decisions are made at the local level, a focus on big-picture issues such as climate change may be lost. For example, in the Pacific Northwest region of the United States, large-scale hydropower produces a significant portion of the electricity. If a single community decides to develop its own microgrid, it is likely that the mix of available energy sources will include coal, natural gas, wood, and other fuels that produce more carbon emissions than hydropower. Finally, opponents argue that citizen engagement is illusory because, just as in other smart grid options, individual participants risk loss of privacy and control. Once the data exist, there is no way to fully guarantee their security.

7.3 Technologies and Actors

The most basic technologies required for a local and community-based energy initiatives are DG, "storage systems, distribution systems, and communication and control systems" (Mariam 2013a). Examples of smart grid technologies that are likely to play central roles in microgrids are rooftop and community-based PV, wind turbines (both individual household and community-level), low-voltage distribution network wires, and electric vehicles (EVs). Although we recognize that some microgrids are based on fossil fuels (such as diesel generation sets used on islands or isolated communities), these systems fall outside the purview of smart microgrids.

One of the most difficult challenges of building small-scale projects is that, with the exception of nanogrids, the number of actors does not decrease as the project becomes smaller. For example, although their roles may shift slightly in small-scale electricity grids, multiple constituents remain involved, albeit in different ways.

A more meaningful way to describe the actors involved in small-scale smart grid initiatives may be to note changes in their roles and relative influence. As more utility customers participate in these initiatives, the utilities "must embrace change in technology and business models in order to maintain a viable utility industry" (Kind 2013). Part of that change is giving up some of their system control (McMahon 2014). This is especially challenging for large, bureaucratically organized incumbents that may have difficulty even imagining different business models. At the same time, they may realize new economic opportunities, so long as the organization is sufficiently nimble to exploit them. But consumers shifting to a prosumer role could gain both additional rights and responsibilities. Must they also accept some of the responsibility that was traditionally held by utilities for managing grid stability? Will they gain the right to influence what energy sources the utility uses to produce electricity?

All of this shifting brings incipient tensions between the different groups of actors, and even within groups of actors, to the surface. For example, the first case we describe in this chapter pits electricity system actors against each other. In this case, a large, traditionally organized utility is engaged in an expensive legal battle with a local government and individual residents of Boulder, Colorado, who are demanding the right to change the mix of resources used to produce their electricity. Civil society organizations are split, with some supporting the utility and others supporting the community. Government entities are also split, with the state's Public Utilities Commission (PUC) supporting the utility against the wishes of the municipal government. As the conflict continues to evolve, different coalitions emerge and cause further splinters in groups that were previously seen as having similar interests. For example, the municipality's proposed changes will provide increased profits for some suppliers of smart grid technologies, which means it is in the best interest of the suppliers to support the municipality rather than the utility, which could have been their traditional ally.

7.4 The Battle in Boulder: Energy Autonomy Through Municipalization

The city of Boulder, Colorado is in the midst of a legal battle with Xcel Energy. A majority of environmentally conscious city residents want to legally end their long-time relationship with Xcel. Rather than remaining dependent on Xcel for the operation of their distribution network, Boulder wants to municipalize its electricity distribution system so that it can control its management. We define *municipalization* "as the process by which municipalities (cities, towns, or counties) take control of the distribution and sometimes generation, of electricity, usually from an investor owned utility" (Browning 2013 pp. 12–13). Municipalization is one response to the United Nations Environmental Programme's call for "governments and local institutions . . . to increase their involvement" in energy systems (United Nations Environment Programme 2012). In this case, Boulder would not take over power generation from Xcel, but would directly manage the low-voltage distribution network and better control the sources of electricity included in the city's power mix. Boulder has ambitious carbon emission reduction aspirations and does not want to continue to accept Xcel Energy's power which relies on coal for over half of its energy gener-ation, though the Colorado Renewable Portfolio Standard requires that 30 percent of its generation comes from renewables by 2020. Currently, Xcel produces over 20 percent of its electricity in Colorado from renewable sources and has the highest percentage of renewable sales of U.S. utilities (Ceres 2014).

The events unfolding in Boulder are indicative of power struggles that are emerging throughout the world. Communities are increasingly attempting to wrest local control over their electricity systems from large, established centralized energy companies. These struggles are part of a new set of demands and expectations with regard to electricity. While electricity system management has traditionally been guided by the need to provide low-cost reliable power, concerns about the environment, climate change, security, and public health have contributed new expectations for the energy system, and fundamentally altered the energy landscape. The Boulder story illustrates tensions that have emerged as part of revamping the electricity system to fulfill expanded social expectations related to enhanced efficiency and sustainability.

7.4.1 Plans for a Smart Grid City

In 2008, media ranging from *MIT Technology Review* to the design-oriented *Inhabitat* trumpeted the news about Boulder's smart grid. *Inhabitat*'s readers learned that Boulder was "poised to become the nation's first fully integrated Smart Grid City" (Trotter 2008). Readers learned from an extensive article in the *Technology Review* that Boulder, Colorado "should soon boast the world's smartest – and thus most efficient – power grid" (Fairley 2008). The *Review* article included information from

Xcel's then chief information officer, who explained that the company had chosen Boulder because it provided an ideal site for experimenting with different approaches to smart grid: a relatively isolated distribution system, with a well-educated and environmentally conscious population. Its size also was large enough to demonstrate how smart grid technologies would work on a commercial scale, without being too large for a controlled experiment. E Magazine proclaimed Boulder "the first to employ the smart grid citywide" (Martin 2010 p. 26). Early results from the project were described as "impressive savings, increased reliability, and excitement for what comes next" (p. 27).

Because media need to tell an exciting story, we sought out information directly from Xcel Energy, thinking that, as the project leader, Xcel Energy could be expected to provide a more balanced if less exciting perspective. Xcel's SmartGridCity webpage reads:

SmartGridCity, in Boulder, Colo., is a fully integrated smart grid community with what is possibly the densest concentration of these emerging technologies to date. It is a comprehensive system that includes a digital, high-speed broadband communication system; upgraded substations, feeders and transformers; smart meters; and Web-based tools available through My Account. Customers that live in this area are now among the first in the world to enjoy a system using smart grid technology to deliver its electricity.

The site promises that Xcel will soon provide Boulder residents with "In-Home device technology evaluation, conservation education, Pricing Plan participant results, plug-in hybrid electric vehicle road tests, and added Web tools" (Xcel Energy 2008). Xcel's news archive from May 2008 includes an announcement that "we're on our way toward building the grid of the future and making SmartGridCity a reality" (Xcel Energy 2008). The news release continues with a detailed description of the project's first phase, which was planned to be conducted between May and August 2008.

However, by the end of 2010, Smart Grid City was embroiled in controversy and widely proclaimed a failure. Early coverage had noted that the project could help smooth tensions that had developed out of Boulder's opposition to Xcel's continued reliance on coal, which represented a barrier to the city's commitments to substantially cut greenhouse gas emissions (Fairley 2008). The project's collapse fueled support for a 2011 referendum to municipalize the city's electrical system, taking control from Xcel (Chediak 2011). When *The Denver Post* investigated the documents and testimony that Xcel was eventually required to file with the Colorado PUC and interviewed available Xcel executives, they found the company had delayed for a year before informing the PUC about projected cost overruns; installed 101 in-home energy devices, rather than the planned 1,850; and abandoned the use of the communications software developed especially for the project (Jaffe 2012). Some of Xcel's partner companies also pulled out of the deal, leaving Xcel short of financial backing. The article in *E Magazine* had hinted at possible financial problems related to installation of the underground cable, quoting an Xcel spokesperson who noted: "when they call it Boulder, there's a reason" (Martin 2010 p. 27).

7.4.2 Municipalization Challenges

Municipally owned utilities (colloquially referred to as munis) provide only a small percentage of U.S. electric power, with approximately 70 percent receiving electric power from private, investor-owned electric utilities, 15 percent from municipal utilities, and another 15 percent from rural electric cooperatives or public power agencies (The Regulatory Assistance Project 2011). Like other companies in the private sector, investor-owned utilities such as Xcel are legally responsible for generating profits for their investors and shareholders. Despite calls to internalize costs associated with fossil fuels such as air and water pollution and direct health effects, many of these costs remain external to current accounting schemes, and do not directly figure into the profits/loss on a company's balance sheet. While Xcel has invested heavily in renewable resources (Ceres 2014), viewed from this perspective, Xcel is fully justified in its continued reliance on coal, which still makes up roughly 56 percent of its Colorado energy generation. Boulder residents are attempting municipalization as a means of changing the fundamental premises that have prevented Xcel from responding to their desire to even more strongly emphasize renewables in its energy generation mix.

Communities that attempt to municipalize face many challenges (Browning 2013). Critics argue that munis have an unfair tax advantage over private utilities, are risky because they cannot diversify their portfolios, lack the large economies of scale that private companies have, and demonstrate government interference in what should be the private sector. Advocates, such as Boulder, respond that munis are more responsive to their consumers, are more able to diversity their energy mix, and provide opportunities for local employment. In Boulder's case, the ability to incorporate more renewables into the energy mix is a primary motivator.

Boulder is a profitable customer; one that Xcel has been unwilling to let go without extracting a high payment to reimburse them for their investments in infrastructure. It made $144 million in gas and electricity sales in Boulder in 2009 (Jaffe 2010). The latest franchise agreement between the city and Xcel expired in 2010, and they began negotiating the new agreement in 2008. Conflicts over the smart grid pilot that Xcel had hoped would resolve some of the city's concerns and opposition to a new coal plant that Xcel began building in 2005 combined to crystalize Boulder's dissatisfaction into a municipalization plan. In 2005, Boulder commissioned a municipalization feasibility study to estimate how much the process would cost the city, but noted that the amount due to Xcel would be determined by a FERC proceeding (Browning 2013). The legal wrangling over relative financial responsibility continues, with Xcel arguing that the city is responsible for millions of dollars in stranded costs and the city arguing that Xcel had no legal right to expect its agreement with Boulder to continue beyond 2010.

Although Boulder's 2011 municipalization referendum passed (as have additional referendums since then), the legal maneuvering continues. In 2013, Xcel's Director of IT Infrastructure and Smart Grid maintained "that Smart Grid City was completed successfully, but he acknowledged how the company failed to educate its customers" (Nowicki 2013). Xcel argued that the proposed municipalization should not be allowed because it "could lead to spiraling costs" for electricity consumers (Nowicki 2013). Not surprisingly, studies commissioned by Boulder tend to show that financial costs of municipalization are manageable, and emphasize the likelihood of environmental and reliability benefits (Beck 2005; Robertson-Bryan 2011). Studies commissioned by Xcel, on the other hand, tend to show that financial aspects of municipalization impose an unfairly high cost on consumers, and emphasize the uncertainty of environmental and reliability benefits (Utilipoint International Inc. 2011).

In her review of municipalization attempts, Browning identified factors that had contributed to past successes (Browning 2013). She found that the Boulder case exhibited six favorable factors:

the local government is on the whole well-perceived by Boulder residents, funding was available for the campaign effort (though Boulder did not outspend Xcel by any means, enough was spent that the referendums passed), the incumbent utility is unpopular in the community (at least to some extent), the community has done substantial research on municipalization, and there is a well-articulated reason for municipalizing (Browning 2013 pp. 92–93).

At the same time, municipalization of Boulder's electricity system is constrained by the complex network of policies and regulations ranging from state to federal-level legislation. In 2014, for example, the city appealed two rulings of the Colorado PUC, which had supported Xcel's position (City of Boulder 2014a).

7.4.3 Building a Smart Municipal Electric Grid

Boulder's attempt to municipalize its electricity is part of its sustained effort to maximize the percentage of energy provided by renewable resources. The city had been pressuring Xcel to increase the percentage of renewable energy in its mix since at least 2005, viewing Xcel's continued reliance on coal as a barrier to the city's commitment to substantially cut its greenhouse gas emissions. It responded to Xcel's claim that its demands are unfeasible by pointing to research such as the 2012 National Renewable Energy Laboratory (NREL) study that showed the United States could use renewable resources to supply 80 percent of its electricity needs by

2050, even if limited to technologies that were commercially available in 2012 (National Renewable Energy Laboratory 2012). Although the Smart Grid City debacle extenuated the already fraught relationship between Boulder and Xcel, it did not lead Boulder to reject the technologies and concept of smart grid. Instead, the city decided it could do a better job on its own.

Boulder's website includes a detailed Climate Action Home Page, with the assertive headline "Let's show the world how it's done" (City of Boulder 2014b). Boulder's first substantive program designed to reduce greenhouse gas emissions was the 2002 Climate Action Plan (CAP). In 2006 the city added the CAP tax, describing it as "the nation's first tax exclusively designated for climate change mitigation." In 2012, voters approved a renewal of the CAP tax. The CAP now includes an outline of Boulder's work plan leading to the eventual operation of its own electric system. Boulder was awarded a 2013 Climate Leadership Award, with "its goals, implementation strategy, and stakeholder engagement" recommended as "a model" for other U.S. locations (U.S. Environmental Protection Agency 2014). Municipalization of the city's electricity forms a central component of the award-winning strategy, however, and its feasibility remains uncertain. The city has approved ordinances which allocated $214 million to negotiate purchase of Xcel's electricity system assets, including the substations, distribution lines, and other infrastructure (City of Boulder 2014a). This is a legally required step before the city can file for condemnation of the electricity system which serves Boulder citizens. However, some of the property is outside of Boulder city limits, and may serve county residents. The Colorado PUC has argued that *they*, not the City of Boulder, have the right to approve the condemnation of the electricity system. The City of Boulder disagrees and the matter is under review (City of Boulder 2014a). Condemnation cannot proceed until the Federal Energy Regulatory Commission (FERC) determines the value of the assets being condemned; a determination that is likely to take several years.

At the time of writing, the city still has initial implementation of its new municipal utility slotted for 2016. Boulder residents have demonstrated the motivation to configure their own community energy system, with smart grid as a key component. At this point, we cannot predict whether Boulder will be allowed to or afford to municipalize (or how successful their effort will be). Still, when attempting to create a municipal electric grid that boasts "Clean Local Energy," it probably helps to be listed as the most educated city in the United States (Kurtzleben 2011; City of Boulder 2014c).

The ongoing battle for the Boulder electricity system shows how difficult it is for any community to gain control of its electricity system. The process is dependent on state laws allowing municipalization, the implementation of that policy by state agencies such as the PUC, approval by federal agencies such as FERC, and the ability of the municipality to buy out the existing franchisee. The relative economic prosperity, educational level, and political acumen of Boulder's residents have enabled the city to fund the necessary studies, understand the results of these studies, and begin to navigate the complicated politics at state, regional, and federal levels. Those characteristics also make Boulder a customer that Xcel is reluctant to give up; and beyond the direct costs of losing Boulder, indirect costs also loom in terms of image and precedence for similar actions by other municipalities.

7.5 Pecan Street Incorporated

The Pecan Street Project (or Pecan Street), in Austin, Texas, shares many characteristics with the smart grid project envisioned for Boulder, Colorado. At the same time, it differs in important ways. If the battle for Boulder demonstrates how a local community of "energy rebels" (Krause 2013) tries to promote change by taking over the system, Pecan Street demonstrates how a local community tries to promote change by working within the system.

Pecan Street is a Regional Demonstration Project funded primarily through the American Recovery and Reinvestment Act of 2009, which provided resources to enable implementation of Title XIII of the Energy Independence and Security Act of 2007 (Austin Energy 2014). The project has been deployed within a 711-acre mixed-use development built on the site of a former DoD installation. The project integrates home energy monitoring systems, a smart meter research network, energy management gateways, distributed generation like solar PV, electric vehicles, and smart thermostats. This technology assemblage forms a smart microgrid that links 1,000 residences, twenty-five commercial properties, and three public schools.

As with Boulder, the initiative for the Pecan Street Project grew out of local demands and expectations placed on the electricity industry. Austin has a Climate Program that began nearly a decade ago. In 2007, Austin City Council passed a resolution to establish a Climate Protection Plan for the purpose of significantly reducing the city's greenhouse gas emissions. In 2011, the Council approved the Austin Energy Resource, Generation, and Climate Protection Plan, which updated goals to more aggressively mitigate emissions through 2020 (Austin Energy 2014). With stereotypically Texan flamboyance, the program's stated goal is to "make Austin the leading city in the nation in the fight against climate change" (Austin Energy 2014). Like Boulder, Austin received a Climate Leadership Award in 2013, with specific recognition for "tracking comprehensive GHG inventories and for its progress on aggressive emissions reduction goals" (Gregor 2013).

7.5.1 Building on Existing Identities and Political Infrastructure

Unlike Boulder, however, Austin has long been served by a municipal electric utility that it has been pressing to move beyond providing low-cost reliable power to responding to environmental concerns, especially with regard to climate change. As a municipal utility, Austin Energy is directly responsible to the Austin City Council, and is highly motivated to be responsive to citizen concerns. It boasts that its "GreenChoice® program is the nation's most successful utility-sponsored and voluntary green-pricing energy program" (Austin Energy 2014). Austin Energy is the eighth-largest publicly owned utility in the U.S., and provides electrical power to approximately 450,000 customers across approximately 450 square miles in central Texas. As a department within the city, it returns profits to help fund other services ranging from police to libraries (Austin Energy 2014).

The Pecan Street Demonstration Project began in 2008 as an effort to exploit the historical connections between Texas and the energy industry, as well as the emerging IT capabilities and environmental consciousness of Austin (Pecan Street Inc. 2013d). Texans view Austin as the state's progressive and liberal city (Feldpausch-Parker et al. 2009). Austin has one of the oldest Green Builder programs in the United States and supports a suite of Smart Growth policies, which include a number of environmentally friendly incentives and initiatives for energy and water conservation.

With its demonstrated commitment to climate change mitigation, Austin symbolizes the expansion of the Texas identity from the leading U.S. *oil*-producing state to the leading U.S. *energy*-producing state. Austinites pride themselves on their outstanding performance as participants in national and international energy programs that are influencing the future of energy production, transmission, delivery, and use. Along with its progressive persona, Austin continues to exude the frontier spirit of adventure, expansion, independence, and limitless possibility that makes up the Texas mythology. Besides being the Texas state capital, it is the home of the University of Texas, Dell Computers, Whole Foods, and Willie Nelson (Feldpausch-Parker et al. 2009).

In 2008, the City of Austin (including, but not limited to, Austin Energy) began discussions with the University of Texas, the Austin Technology Incubator, and Austin's Chamber of Commerce. By the end of the year, it had begun recruiting corporate partners (Pecan Street Inc. 2013a). In 2009 an expanded group began researching possible projects and formed the Pecan Street Project Inc. as a nonprofit organization to coordinate their efforts. The new organization applied for Department of Energy funding for a smart grid demonstration project; it was awarded $10.4 million. Planning for implementation and continued recruitment of corporate partners occupied most of 2010, and implementation began in 2011. The word "project" was dropped from the organization's name as it expanded beyond the original vision of a single smart grid demonstration project in Austin (Pecan Street Inc. 2013b). Pecan Street Inc. has split into two segments, with one focused on smart grid research and the other on commercialization of smart grid technologies (Pecan Street Inc. 2013b). In 2014 the research institute launched WikiEnergy, which offers "the world's largest research database of customer energy and water use" (Pecan Street Inc. 2014). The consortium is expanding into California, with a new development in San Diego's Mission Valley called Civita (Pecan Street Inc. 2013c).

7.5.2 Smart Grid as Texas-style Adventure

While the celebratory rhetoric of Pecan Street Inc.'s promoters is to be expected, what about perspectives from beyond its local developers and commercial partners? Prior to the project launch, the media buzz was similar to that for the proposed Boulder project. For example, when describing the project, *Inhabitat* quoted an unnamed spokesperson from Austin Energy as claiming that the project would be part of transforming the city into "the urban power system of the future while making the City of Austin and its local partners a local clean energy laboratory and hub for the world's emerging cleantech sector" (Schwartz 2009). Two years later, *Smart Grid News* described the Pecan Street Project as providing the "sizzle," "Disney-style magic," and "razzle-dazzle" needed to spark public interest in smart grid (Enbysk 2011). The article went on to describe the exciting household appliances, consumer electronics, and electric cars that were included in the Pecan Street package due to partnerships with commercial giants such as Sony, Intel, Whirlpool, Chevrolet, Landis+Gyr, and Best Buy.

The Environmental Defense Fund (EDF) has staked its credibility on backing Pecan Street. The same article in *E Magazine* that featured Boulder as the nation's first citywide smart grid also introduced Pecan Street as an innovative partnership between local government, nongovernmental organizations, and the private sector in the form of a varied group of commercial organizations (Martin 2010). The author quoted James Marsdon, EDF's Energy Program Director, stating that "EDF views Pecan Street as so groundbreaking, we believe we'll be able to recommend it as a future model for other utilities around the nation" (Martin 2010 p. 29). With climate and energy as one of its focus areas, EDF features Pecan Street on its website, where it proclaims that Pecan Street "is a laboratory of ideas and technologies that will move the nation's $1.3 trillion electricity market toward a future in which energy is cheap, abundant and clean. If Pecan Street is successful, every neighborhood in America will look like it in 20 years" (Environmental Defense Fund 2013). And when U.S. Department of Energy Secretary Ernest Moniz visited in 2014, he declared that Pecan Street Inc. was a "vibrant, innovative ecosystem" for energy development (Price 2014).

7.6 Denmark's Ecogrid: Bliss in the Baltic

The European Union (EU) has demonstrated a strong interest in developing smart microgrids, as demonstrated by an innovative project located on the Danish island of Bornholm. Bornholm, which hosts the EU's first full-scale deployment of EcoGrid EU, is the site of a smart microgrid that has introduced "market-based mechanisms close to the operation of the power system" (Lohse 2014). Locating Ecogrid on Bornholm has enabled the EU to literalize the island metaphor and conduct realistic testing of a novel approach to electricity that it hopes will provide a model for smart grid development throughout the world. In the understated style of project coordinator Ove S. Grande explains, their "hope is that the experiences from EcoGrid EU will contribute to ... the development of the European 20–20–20 energy and climate goals" (EcoGrid EU 2013a p. 3).

Bornholm lies in the Baltic Sea, 200 kilometers to the east of Copenhagen. When the island comes up in conversation, Danes "tend to go misty eyed... This island, with its time warp, red-roofed fishing villages and magical beech forests holds a special place in the national collective memory. All will have made the pilgrimage there at least once, usually first as part of a school trip, and then perhaps a second or third time with families" (Booth 2013). According to legend, when the gods created the world, they saved all the best bits for last, kneaded them together, and then tossed them into the Baltic to form Bornholm (Kumagai 2013).

Bornholm has capitalized on its thoroughly documented status as the sunniest spot in Denmark and hosts about 600,000 tourists each year, many of them returning Danes. It is famed for its dramatic rock formations in the north that slope down into lush forests and its magnificent sandy beaches, primarily in the south. Its unique natural features, historical ruins, locally produced food specialties, and ceramics and glassware artisans make it a popular tourist destination, especially for Scandinavians, Germans, and in recent years Poles (Miljøministeriet – Danish Forest and Nature Agency 2013; Velkomstcenter 2013).

Of course, the island also has a permanent population. Ronne, the main town, has a population of 13,000. Ronne, along with other villages ranging in size from 70 to 4000 persons, gives the island a population of approximately 45,000. This permanent population works in a broad range of small enterprises, providing "a representative mixture of commercial, industrial, and residential customers, as well as schools, a hospital, an airport, and an international seaport" (Kumagai 2013). The advantages Bornholm offers for developing a smart microgrid are similar to those Xcel found in Boulder: the distribution system can be isolated, the population is well-educated and environmentally conscious, the size is large enough to pilot-test smart grid technologies at scale. But unlike Boulder, Bornholm has its own municipal utility. Maja Bendtsen, an engineer with Østkraft, the island's municipal utility, explained that because Bornholm is "a microcosm of Danish society," it facilitates realistic study of multiple ways to reach the EU's goals of cutting greenhouse gas emissions 20 percent by the year 2020. For Jacob Østergaard, a professor at the Technical University of Denmark, the underground cable is what makes Bornholm an ideal site for Ecogrid EU, because he can switch the cable on and off to conduct realistic field experiments (Kumagai 2013).

7.6.1 Smart Grid as Danish Design

Given the proverb "necessity is the mother of invention," Bornholm and Denmark may provide the perfect seedbed for designing smart microgrids. Although Denmark has access to oil from the North Sea and is an oil exporter, all coal is imported. In an attempt to reduce dependence on coal and other fossil fuels, Denmark has developed a plan to completely phase out fossil fuel use by 2050 (Danish Energy Agency 2013; Danish Ministry of Climate Energy and Building 2013). In 2012, wind power was already providing more than 30 percent of the electricity consumed in Denmark, with a goal of 50 percent by 2020. For many years, Bornholm has obtained its electric power via an undersea power cable that connects it with the Nordic grid (Kumagai 2013). The cable was accidentally severed four times in the ten years prior to launching Ecogrid EU, which may have contributed to Bornholm's readiness to host Ecogrid.

Østkraft had already begun introducing distributed energy resources (DERs) into the Bornholm system, both to enable rapid response when the undersea power cable was accidentally damaged and to facilitate integration of the large amount of wind power that is locally available (Kumagai 2012). The Ecogrid project gave Bornholm the financial resources to develop a virtual power plant that aggregates all of those DERs. Although ten years ago virtual power plants were mostly experiments, the big players in energy markets have begun to recognize them as a commercially viable alternative to building traditional power plants to cover peak capacity needs. Most allow large electricity customers (that is, industrial production sites) to trade energy in the day-ahead market. Kim Behnke, who heads R&D at the Danish utility Energinet, describes that process as "Smart Grid, version 1," noting that on Bornholm "we are going for Smart Grid, version 2" (Kumagai 2012).

Ecogrid's project leaders have collaborated with many of Bornholm's 45,000 permanent residents to design a smart microgrid that demonstrates that quintessentially Danish term, *hygge*. *Hygge* translates roughly into English as "coziness." Along with one's physical surroundings, *hygge* includes a social dynamic. Although achieving *hygge* may involve design elements connected to a nostalgic view of the past (such as soft lighting produced by flickering candle flames), it does not embrace a naïve belief that nature automatically produces coziness. *Hygge*, or "the art of creating intimacy: a sense of comradeship, conviviality, and contentment rolled into one," includes an explicit awareness of design (Fathom 2011). And the entire Ecogrid project was designed with *hygge* in mind. In 2012, recruitment kicked off at a giant block party in the demonstration house, Villa Smart (EcoGrid EU 2013b pp. 24–25). During the next few months, printed materials were handed out at social events, including Bornholm's annual Energy Day. Approximately 1,000 participants signed up during the first year. A formal campaign to double this number was designed around more social events, all with live entertainment and plenty of food and drink, as well as direct mailings. By August 2013, the project had its required 1,900 household volunteers, just under 100 commercial and industrial volunteers, and a waiting list (EcoGrid EU 2013b).

One important aspect of this project is that it "appealed more to good citizenship rather than narrow financial gains" (EcoGrid EU 2013b p. 32). Recruitment material focused on social and environmental values, emphasizing that participants would be helping develop an electrical system that contributes to the Danish goal of replacing fossil fuels with local, renewable energy, as well as the global need to mitigate anthropogenic climate change. Although volunteers were promised that their participation would not result in additional costs, they were not promised that they would save money.

According to Dieter Gantenbein, a smart grid researcher with IBM-Zurich, this approach fits well with the Danish social consciousness. "Danes take preservation of the environment close to their hearts... It's like a sport... They use different technologies, and by being engaged, they are very enthusiastic to participate in such an ambitious pilot" (Kumagai 2013 p. 6). Martin Kok-Hansen, a real estate agent in Ronne, illustrates this heightened awareness when explaining how his smart equipment guides his decisions that help to balance the grid. Since heating is one of the biggest energy demands, he notes that if the utility needs the power, he can allow it to switch off his heat until the temperature is "two or three degrees cooler than normal. . . Maybe you put on a sweater for a while" (p. 6).

Not everyone living on Bornholm is a smart grid enthusiast and/or an environmentalist, and Ecogrid's designers realized they needed to involve those who are not enthusiasts, both to obtain sufficient numbers for the test and to support claims about the project's generalizability. Recruiters found that people who were not especially interested in energy or the environment got excited about the equipment – one of the few volunteers who dropped out of the project did so "because he did not like the design of the smart boxes" (EcoGrid EU 2013a). And grocery store operators assume that, although most customers are unaware that the refrigerators keeping the beer cool are programed to detect conditions on the grid and then switch on and off as needed to balance frequency variations, they would be pleased if they heard about the innovation (Kumagai 2013).

7.7 Nanogrids to the Rescue

As we noted at the beginning of this chapter, nanogrids are the smallest and simplest of the microgrids. They may be nothing more than a load and a gateway, but, as Bruce Nordman notes, "the fact that they are small and simple does not mean they are not useful and important" (Nordman 2010 p. 1). A nanogrid can power a car, a smart building, a campus, or a remote village (Nordman 2010; Nordman et al. 2012; Hardesty 2014). Nanogrids simultaneously provide a more conventional and more radical approach to electricity than other microgrids. They are more conventional than other microgrids because "they do not directly challenge utilities" (Hardesty 2014). Perhaps because they seem relatively nonthreatening to utilities and other legacy actors, nanogrids have not been opposed with the same arguments about "real or imagined complexity" that other microgrids have encountered (Nordman 2010). Peter Asmus, the person who popularized (and perhaps invented) the word "locavolt," claims that "nanogrids represent a larger market opportunity because they are . . . less challenging to the status quo and less subject to the technical challenges facing larger distribution networks" than other microgrids (Lundin 2014b).

At the same time, nanogrids are technologically radical in their tendency to rely on direct current (DC) and could be politically radical in their reversal of the top-down approach that characterizes the energy system (Asmus and Lawrence 2014). Using DC minimizes conversion losses, and working directly with consumers could help to maximize the likelihood that the system responds to local needs. These modular building blocks can support a wide variety of energy applications that are relevant to consumers ranging from the U.S. DoD to communities that have no access to electricity. Navigant Research has forecast that nanogrid vendors will generate $59.5 billion by 2025 (Asmus and Lawrence 2014; Lundin 2014b,). They cite the increased integration of electric vehicles to provide emergency electricity or storage to reduce peak demand from buildings as an illustration of one of the more radical applications of nanogrids.

The DoD, which is the largest energy consumer in the United States (Chen 2012a) and has strong motivation to ensure energy security for its bases and field operations, is testing nanogrids on several bases. Its nanogrid tests range from a building-integrated photovoltaic roof at the Marine Corps Station in Yuma, Arizona to a plug-in vehicle demonstration at the Los Angeles Air Force Base (Chen 2012a). Both of these projects illustrate the flexibility of nanogrids. The plug-in vehicle and the building can operate independently of the larger system, and also can be integrated into the larger system when such integration is useful. The DoD's SPIDERS program demonstrates a more programmatic approach, and focuses primarily on protecting "critical infrastructure from power loss in the event of physical or cyber disruptions" (Sandia National Laboratories 2012 p. 1). The plan is to design and develop three increasingly complex microgrids that would allow the sites to maintain critical operations if the regular electricity supply were interrupted. The first SPIDERS installation will demonstrate the simplicity that characterizes nanogrids, with the second and third gradually integrating a more complex portfolio into each microgrid.

While nanogrids can operate independently, they can also connect with microgrids and the larger grid. When Superstorm Sandy knocked out power in the northeastern United States, Princeton University's energy manager, Ted Borer, "flipped switches that restored power to much of the campus" (Clayton 2012). Although Princeton does not usually operate in an islanded mode, the ability to do so meant it could continue functioning, although at a reduced capacity. Other nanogrids also helped mitigate the immediate consequences of Sandy. The Federal Drug Administration's White Oak facility in Maryland maintained power in all of its buildings, New York University provided power and heat to a portion of its Manhattan campus, and South Windsor High School in Connecticut used its nanogrid to power an emergency shelter (Clayton 2012).

Given all of their proven functionality, the growing interest in nanogrids is not surprising, and they are "expected to grow significantly in number, usefulness, total energy distributed (Nordman 2010). The likelihood of nanogrids achieving their potential can be increased by defining a standard architecture, providing a single specification for gateways, keeping power distribution and functional control separate, and testing or demonstrating possible outcomes of nanogrid connections. Because nanogrids have demonstrated that they can be complementary to conventional electric sector business models, utilities and other legacy actors have remained relatively agnostic toward their deployment. However, this could change if nanogrids enable significant expansion of DG and cut into utility profits. For the time being, the overwhelming appeal of nanogrids is the varied applications of smart grid they enable, while presenting an apparently minimal threat to the existing system.

7.8 Highlighting Smart Grid's Transformative Potential

Small-scale grid innovation highlights the transformative potential of smart grid. According to Jeremy Rifkin, an economic and social theorist who is equally at home in higher education, business, and government, "Internet technology and renewable energy are merging to create a powerful 'Third Industrial Revolution'" (Rifkin 2014). This revolution brings dramatic changes to the dominant paradigm of the twentieth century, with its top-down approach to management and its assumption that large, centralized organizations are the most economically feasible (Wright 2012). Although some dismiss Rifkin and his talk of revolution as visionary and unrealistic, he has served as an advisor for several businesses and government leaders, and is currently an advisor to the EU. His advice regarding smart grids is to focus on local control. The public–private partnerships that have emerged out of both the Pecan Street Project and Bornholm's Ecogrid EU illustrate the social and economic capital that this approach can generate.

This revolution includes a shift away from maintaining the separation between those who produce electricity and those who consume it. Similar to the content creators reshaping digital media, smart grid could enable more direct engagement with the energy system. Rather than centralized producers building the right energy system for consumers, small-scale smart grid can empower prosumers to decide for themselves what energy system is right, and then how to develop that system. Many of these concepts are not new, but link back to a 1970's ethos that questions unsustainable economic growth and large-scale energy systems while promoting self-sufficiency (Daly 1973; Lovins 1977,).

As we described in Chapter 2, becoming a prosumer involves much more than putting up a wind turbine and/or installing solar panels on the roof. Prosumers could also be empowered to change the rules, which include both cultural norms and legal frameworks. Because their stake in the energy system is relatively new, they have little enthusiasm for maintaining structures that appear to be reproducing negative results. They are free to imagine an electricity system that fulfills their needs, and even their dreams.

All three of the stories we reviewed in this chapter demonstrate how smart grid development is being used to spark innovation and expand energy markets. For example, despite the public challenges faced by Xcel's Smart Grid City, the experiment provided valuable information to the utility. The technologies it tried out enabled it to better manage system voltage, save fuel, and cut customer complaints about power fluctuations (Jaffe 2012). And the ongoing struggle between Xcel and Boulder has led to other conversations about the role of communities in power system management and new initiatives (Cardwell 2013; Xcel Energy 2014).

The original Pecan Street Project has become Pecan Street, Inc., which hosts Pike Powers Lab, where smart grid technologies are tested, and a Wiki that includes the world's "largest residential energy database" (Pecan Street Inc. 2013b; Pecan Street Inc. 2014). The partnership, which is headquartered at the University of Texas, has boosted the university's profile in energy research (University of Texas 2014) and has recently launched another real estate development in San Diego, California (Pecan Street Inc. 2013c). The Bornholm project has provided similar benefits for the Technical University of Denmark, which has emerged as an international leader in smart grid research on topics such as distributed generation, control schemes, customer engagement, scalability, and virtual power plants (Technical University of Denmark 2014). It has provided opportunities for collaborative field experiments, where Østkraft (the island's municipal utility) works with private companies ranging from corporate giants such as IBM and Siemens to local grocery markets to test different technological configurations for managing distributed energy sources, including novel ideas for storage (Østergaard and Nielsen 2011).

Examining small-scale smart grid initiatives also highlights the power of the prosumer. Although not all electricity consumers want to become prosumers, small-scale initiatives maximize the opportunities for those who do want to assume a more active role. This includes both hightech locavolts such as the politically active residents in the three communities we discussed in this chapter, and people living in remote locations in the Australian Outback or the interior of Alaska. The U.S. military's interest in becoming an active participant in grid modernization illustrates yet another type of actor. Despite its large financial portfolio and powerful political influence, the U.S. military generally purchased power from the local utility, and accepted the passive role of consumer when it came to electricity. With the advent of smart microgrids it has strategically embraced the role of prosumer, including a reimagination of what services and tactical advantages the electric grid can and should provide (Perera 2012).

These initiatives demonstrate how smart grid may provide electricity for people who currently have no access. Researchers at Lawrence Berkeley National

Laboratory argue that one of the most important reasons to develop smart grid has been largely ignored (Nordman et al. 2012): smart nanogrids, the smallest of microgrids, are uniquely positioned to enable leapfrogging over much of the electricity system that has dominated twentieth-century development, because they can supply local needs "at a lower cost and reduced energy use" (Chen 2012b p. 26). Although initial implementation may seem expensive, that expense needs to be contextualized. Diesel and kerosene, upon which people often rely when they have no access to electricity, are expensive in terms of both direct and indirect costs, such as their contributions to air and water pollution and health impacts (Tweed 2013). In these situations, electricity can enable local residents to use locally available resources to dramatically boost their standard of living and quality of life.

Finally, small-scale initiatives highlight the flexibility of smart grid development, suggesting the possibility of multiple smart grids that are purpose-designed and tailored to the user, both at the level of individual technologies and at the system level. Many of the individual technologies used when deploying smart grids can be configured differently depending on the needs of the user. For example, participants in Bornholm's EcoGrid may decide to have their home heating automatically curtailed if the system needs help in balancing overall frequency, but that decision comes with all sorts of conditions and there is no requirement that all residents make the same decision. There is room for variation in how long the curtailment may last, how much temperature variation is allowed, and other customer-specific adaptations. Participants in the Pecan Street Project might decide that, although they are willing to adjust the time of day at which their dishwasher runs, they are not willing to curtail air conditioning on hot summer days. And participants in both projects can use their household equipment to learn where they are using the most electricity, and identify appliances that may be operating incorrectly (and using more electricity than needed). Of course, small-scale approaches also draw attention to the risks associated with smart grid. In an article that urges both maintaining and building more traditional (fossil fuel-reliant) baseload units, former power company CEO and climate change advocate Charles Bayless argues the contemporary electricity system in North America is among the most "reliable systems in the world, and is a product of billions of dollars of investment and careful planning" (Bayless 2010 p. 75). He urges caution in making fundamental changes to a system that works most of the time for most users, and claims that relying on renewable sources of energy threatens the system's reliability and stability. He suggests that, although smart grid technologies may minimize these problems, they also add unnecessary complexity to the system. In the current system, he claims, "four 1,000-MW elephants pull your system in the same direction, [but in a smart grid system] a thousand 4-MW cats pull in different directions" (Bayless 2010 p. 81). The varied operations that function simultaneously in the small-scale projects described in this chapter highlight this complexity.

The same flexibility that enables participants in small-scale smart grid initiatives to make individual choices also can exacerbate the already widening gulf between the haves and the have-nots. As Xcel has argued, if the utility loses the relatively well-off city of Boulder, that places more economic stress on less prosperous communities within its service territory. In Arizona, where use of solar power is rapidly expanding, households that have installed rooftop solar panels generate their own electricity most of the time, and rely on the publicly available grid when they need it (Brandt 2013). Because they use the larger grid at little or no cost, they have an unfair advantage over those who cannot install rooftop solar panels, either because they do not own their homes or because they simply cannot afford to purchase and install the panels. Although the Arizona Public Service Company (Arizona's PUC) has submitted a recommendation for regulatory reform to correct this inequity, the situation illustrates the importance of attending to the details of how smart grid is deployed and how associated regulations develop.

These examples also highlight how important it is for law and policy to keep up with technological change. Oleg Logvinov, who serves on the IEEE-SA Standards Board and Corporate Advisory Group, identified smart grid as "a core subset of IoT [the Internet of Things]," which he expects to fundamentally change everything about society (Lundin 2014a). Loginov explained that "all the elements of smart grid, from generation – centralized or distributed – to transmission to distribution, where the real action is, will be nodes on the IoT," although regulatory and standardization frameworks still need to be rationalized (Lundin 2014a). This is similar to Rifkin's argument that the Third Industrial Revolution, which he sees traveling along the IoT and relying on energy produced by thousands of horizontally organized grids, requires updating regulatory frameworks to enable seamless integration (Rifkin 2014). One barrier to that revolution may be lack of interoperability; the individual technologies and machines need to communicate with each other, and they need to do so instantaneously. Geoff Brown, CEO of Machine-to-Machine Intelligence Corporation, argues that, just as machine-to-machine communication and a functional IoT are essential for successful grid modernization, smart grid is essential to a functional IoT (Brown 2014). He suggests implementation of new messaging protocols like the MQ Telemetry Transport as a means of assuring interoperability among the many individual technologies, and points to NIST's recently released report on how to improve the security of critical infrastructure as a positive sign that policy is catching up with technology (National Institute of Standards and Technology 2014).

Who controls the community energy system, microgrid, and nanogrid and who benefits from these different systems remains context-specific. How these small-scale grid technologies will evolve over time and if they allow for the creation of prosumers or embed the wishes of the locavolt movement remains to be seen. The technical evolution of small-scale energy systems will be supported or thwarted by the different regulatory and policy contexts within which they are developed.

References

Asmus, P. (2008a) Here Come the "'Locavolts!". www.peterasmus.com/journal/2008/7/20/here-come-the-locavolts.html

Asmus, P. (2008b) Think Globally, Power up Locally. *San Francisco Chronicle*. www.sfgate.com/green/article/Think-globally-power-up-locally-3269357.php

Asmus, P. and M. Lawrence. (2014) *Nanogrids: Grid-Tied and Remote Commercial, Residential, and Mobile Distribution Networks: Global Market Analysis and Forecasts.* Boulder, CO: Navigant Consulting, Inc.

Austin Energy. (2014) *Austin Energy: More than Just Electricity.* Austin, TX: Austin Energy.

Bayless, C. E. (2010) The Case for Baseload. *Electric Perspectives*, September/October 2010, 72–83. www.mydigimag.rrd.com/publication/?i=47255

Beck, R. W. (2005) Preliminary Municipalization Feasibility Study. Boulder, CO. www-static.bouldercolorado.gov/docs/energy_future_2005_Preliminary_feasibility_study_from_RWBeck-1-201306061215.pdf

Booth, M. (2013) Denmark: Bornholm, A Danish Paradise. *The Telegraph*. www.telegraph.co.uk/travel/destinations/europe/denmark/10353634/Denmark-Bornholm-a-Danish-paradise.html

Brandt, D. (2013) As State's Solar Use Grows, Electric Grid Access Must be Fair. *AZ Central*. www.azcentral.com/opinions/articles/20130719arizona-solar-electric-grid-access.html

Brown, G. (2014) Smart Grid Interoperability: We Now Have an Answer, Its Initials are MQTT. *Smart Grid News.com*. www.smartgridnews.com/artman/publish/Technologies_Standards/Smart-grid-interoperability-We-now-have-an-answer-its-initials-are-MQTT-6386.html#.U3oRYZ1vnct

Browning, K. C. (2013) *Electric Municipalization in the City of Boulder: Successful Greening or Path to Bankruptcy?*, 92. Claremont, CA: Claremont McKenna College. Senior Thesis. scholarship.claremont.edu/cgi/viewcontent.cgi?article=1711&context=cmc_theses

California Energy Commission. (2009) *A Step by Step Tool Kit for Local Governments to Go Solar*. California Energy Commission. www.energy.ca.gov/2009publications/CEC-180-2009-005/CEC-180-2009-005.PDF

Cardwell, D. (2013, March 13) Cities Weigh Taking Over From Private Utilities. *New York Times*. www.nytimes.com/2013/03/14/business/energy-environment/cities-weigh-taking-electricity-business-from-private-utilities.html?pagewanted=all

Carvalho, A. and Tarla Rai Peterson. 2012. *Climate Change Politics: Communication and Public Engagement*. Amherst, NY: Cambria.

Ceres. (2014). Benchmarking Utility Clean Energy Development. www.ceres.org/resources/reports/benchmarking-utility-clean-energy-deployment-2014

Chediak, M. (2011) Boulder Finds 'Smart Grid' Slow, Pricey. *Bloomberg News*. http://www.bloomberg.com/news/2011-11-01/boulder-finds-smart-grid-slow-pricey.html

Chen, A. (2012a) Berkeley Lab Continues to Support the U.S. Department of Defense in Promoting Efficiency and National Security. *Environmental Energy Technologies Division News*, 11, 11–16. www.eetd.lbl.gov/newsletter/nl41/pdf/eetd-nl41.pdf

Chen, A. (2012b) Nanogrids Can Support Smart Grid Success. *Environmental Energy Technologies Division News*, 11, 26–27. www.eetd.lbl.gov/newsletter/nl41/pdf/eetd-nl41.pdf

City of Boulder. (2014a) *City's Appeal with District Court on PUC Filing Boulder, Colorado.*

City of Boulder. (2014b) *Climate Action Home Page.* Boulder, CO: City of Boulder. bouldercolorado.gov/climate/climate

City of Boulder. (2014c) *Energy Future and the Municipalization Exploration Project.* Boulder, CO. bouldercolorado.gov/energy-future

Clayton, M. (2012) Lessons from Sandy: How One Community Kept the Lights On. *Christian Science Monitor.* www.csmonitor.com/USA/2012/1115/Lessons-from-Sandy-how-one-community-in-storm-s-path-kept-lights-on

Daly, H. E. (1973) *Toward a Steady-state Economy.* San Francisco, CA: WH Freeman.

Danish Energy Agency. (2013) *Our Future Energy.* www.ens.dk/sites/ens.dk/files/policy/danish-climate-energy-policy/our_future_energy.pdf

Danish Ministry of Climate Energy and Building. (2013) *Energy Policy Report 2013.* http://www.ens.dk/sites/ens.dk/files/policy/danish-climate-energy-policy/dkenergypolicyreport2013_final.pdf

EcoGrid EU. (2013a) Eco Grid EU. www.eu-ecogrid.net

EcoGrid EU. (2013b) Eco Grid EU: From Design to Implementation. European Union's Seventh Framework Programme for Research. www.zurich.ibm.com/pdf/ecogrid/131004_%20edk%20a4_ecogrid%20eu%20project_web.pdf

Enbysk, L. (2011) Can Pecan Street Deliver Some Smart Grid Sizzle? *Smart Grid News.* www.smartgridnews.com/artman/publish/news/Can-Pecan-Street-deliver-some-smart-grid-sizzle-It-plans-to-try-4088.html#.UzBHyE1OXIU

Endres, D., Sprain, L., Peterson, T. R. (2009) *Social Movement to Address Climate Change: Local Steps for Global Action.* Amherst, NY: Cambria Press.

Environmental Defense Fund. (2013) Living Laboratory Shows How a Smart Grid Works: Austin Community Benefits from Clean Energy System of the Future. www.edf.org/energy/building-smarter-grid-austin-texas

Fairley, P. (2008) A Power Grid Smartens Up: Communications Technologies Will Make Boulder's Grid More Efficient and Environmentally Friendly. *MIT Technology Review.* www.technologyreview.com/news/409775/a-power-grid-smartens-up/

Fathom. (2011) What the Hell is "Hygge"? www.fathomaway.com/postcards/culture/attempt-define-danish-hygge/

Feldpausch-Parker, A., D. Hall, C. Horton, J. Minion, A. Munoz, I. Parker, and T. R. Peterson. 2009. Step it Up in the Lone Star State: How Identity and Myth May Impact a Movement. In *Social Movement to address Climate Change: Local Steps for Global Action*, ed. L. S. D. Endres and T. R. Peterson. Amherst, NY: Cambria Press, 23–51.

Goldman, M. and C. Diaz. (2011) Community Wind Projects May Benefit from Creative Financing Methods. *North American Wind.* www.nawindpower.com/e107_plugins/content/content.php?content.8490

Gregor, K. (2013) City of Austin Wins Climate Leadership Award. www.austintexas.gov/news/city-austin-wins-climate-leadership-award

Hardesty, L. (2014) What's a Nanogrid. *Energy Manager Today.* www.energymanagertoday.com/whats-a-nanogrid-099702/

Independent Press. (2011) Warren and Watchung Schools, Public Buildings to Receive Solar Panels in Somerset County Project. www.nj.com/independentpress/index.ssf/2011/01/warren_and_watchung_schoolspub.html

Jacob Østergaard and J. E. Nielsen (2011) The Bornholm Power System: An Overview. *DKU White Papers.* www.ctt.sitecore.dtu.dk/upload/sites/powerlabdk/media/the_bornholm_power_system_an_overview_v2.pdf

Jaffe, M. (2010) Boulder Willing to Let Xcel Franchise Lapse while It Studies Future Energy Options. *The Denver Post.* www.denverpost.com/news/ci_16294228

Jaffe, M. (2012) Xcel's SmartGridCity Plan Fails to Connect with Boulder. *The Denver Post.* www.denverpost.com/ci_21871552/xcels-smartgridcity-plan-fails-connect-boulder

Juris, J. (2005) The new digital media and activist networking within anti-corporate globalization movements. *The Annals of the American Academy of Political and Social Science*, 597, 189–208.

Kind, P. (2013) *Disruptive Challenges: Financial Implications and Strategic Responses to a Changing Retail Electric Business*. Washington, DC: Edison Electric Institute. www.eei.org/issuesandpolicy/finance/Documents/disruptivechallenges.pdf

Krause, M. B. (2013) Thousands of German Cities and Villages Looking to Buy Back their Power Grid. *Green Tech Grid*. www.greentechmedia.com/articles/read/Thousands-of-German-Cities-and-Villages-Looking-to-Buy-Back-Their-Power-Gri

Kumagai, J. (2012) Virtual Power Plants, Real Power. *Spectrum*. spectrum.ieee.org/energy/the-smarter-grid/virtual-power-plants-real-power

Kumagai, J. (2013) The Smartest, Greenest Grid: What the Little Danish Island of Bornholm is showing the world about the future of Energy. *Spectrum*. spectrum.ieee.org/energy/the-smarter-grid/the-smartest-greenest-grid

Kurtzleben, D. (2011) The 10 Most Educated U.S. Cities. *U.S. News & World Report*. www.usnews.com/news/articles/2011/08/30/the-10-most-educated-us-cities-boulder-ann-arbor-and-washington-dc-top-the-list

Local Clean Energy Alliance. (2013) Local Clean Energy Alliance: Clean Energy Jobs and Healthy Communities. www.localcleanenergy.org

Lohse, L. L. 2014. References. In *PowerLabDK*, ed. L. L. Lohse. PowerLabDK, Technical University of Denmark. www.powerlab.dk/Access_Use/References

Lovins, A. B. 1977. *Soft Energy Paths: Toward a Durable Peace*. New York: Harper Colophon.

Lundin, B. V. (2014a) The Horizontal Nature of the Internet of Things. *Fierce Smart Grid*. http://www.fiercesmartgrid.com/story/horizontal-nature-internet-things/2014-02-19

Lundin, B. V. (2014b) Tiny Grids are Big Business. *Fierce Smart Grid*. http://www.fiercesmartgrid.com/story/tiny-grids-are-big-business/2014-03-26

Mariam, L., Malabika Basu, and Michael F. Conlon. (2013a) A Review of Existing Microgrid Architectures. *Journal of Engineering*. http://dx.doi.org/10.1155/2013/937614

Mariam, L., Malabika Basu, and Michael F. Conlon. (2013b) A Review of Existing Microgrid Architectures. *Journal of Engineering*. http://dx.doi.org/10.1155/2013/937614

Martin, M. (2010) The Great Green Grid: A Smart Grid that Lets Us Better Control Our Energy Use May Finally Be Ready to Launch. *E Magazine*, 21, 22–29.

Mayer-Schonberger, V. and Lazer, D. (2006) *Governance and Information Technology*. Cambridge, MA: MIT Press.

McMahon, J. (2014) Steven Chu Solves Utility Companies' Death Spiral. *Forbes*. www.forbes.com/sites/jeffmcmahon/2014/03/21/steven-chu-solves-utility-companies-death-spiral/

Miljøministeriet – Danish Forest and Nature Agency. (2013) *Discover the Nature of Bornholm*. Naturstyrelsen (Danish Ministry of the Environment). bornholmsnatur.naturstyrelsen.dk/English/

Ministry of Foreign Affairs Thailand. (2013) Community Based Renewables: Thai-German Cooperation on Sustainable Energy Development. *Renewable Energy Project Development Programme Southeast Asia*. thailand.ahk.de/fileadmin/ahk_thailand/Events/Renewable_energy/files/130807_Programme_Conference_Community_Based_Renewables_in_Germany_and_Thailand.pdf

Minnesota Department of Commerce. (2013) Community-based Energy Development (C-BED) Tariff. energy.gov/savings/community-based-energy-development-c-bed-tariff

National Institute of Standards and Technology. 2014. Framework for Improving Critical Infrastructure Cybersecurity. www.nist.gov/cyberframework/upload/cybersecurity-framework-021214-final.pdf

National Renewable Energy Laboratory. (2012) *Renewable Electricity Futures Study*. eds. M.M. Hand, S. Baldwin, E. DeMeo, J.M. Reilly, T. Mai, D. Arent, G. Porro, M. Meshek, D. Sandor. Golden, CO: National Renewable Energy Laboratory. www.nrel.gov/analysis/re_futures/, www.nrel.gov/docs/fy13osti/52409-ES.pdf

NAW Staff. (2013) Repower Supplying 24 Community Wind Farms in Germany. *North American Wind*. www.nawindpower.com/e107_plugins/content/content.php?content.12381

NAW Staff. (2014a) GMP Says Kingdom Community Wind Farm Delivers Big Benefits for Vermonters. *North American Wind*. www.nawindpower.com/e107_plugins/content/content.php?content.12530

NAW Staff. (2014b) Siemens Providing Direct-drive Turbines for Community Wind Farm. *North American Wind*. www.nawindpower.com/e107_plugins/content/content.php?content.12595

NAW Staff. (2014c) Vestas Wins 72.6 MW Order for Community Wind Farm in Germany. *North American Wind*. www.nawindpower.com/e107_plugins/content/content.php?content.12683

Nordman, B. (2010) Nanogrids: Evolving Our Electricity Systems From the Bottom Up. *Darnell Green Building Power Forum*. www.nordman.lbl.gov/docs/nano.pdf

Nordman, B., K. Christensen, and A. Meier. (2012) Think Globally, Distribute Power Locally: The Promise of Nanogrids. *IEEE Computer*, September 2012. http://nordman.lbl.gov/docs/nano.pdf

Nowicki, A. (2013) Boulder's Smart Grid Leaves Citizens in the Dark. *Greentech Media*. www.greentechmedia.com/articles/read/Boulders-Smart-Grid-Leaves-Citizens-in-the-Dark

Pecan Street Inc. (2013a) *History*. Austin, TX: Pecan Street Inc.

Pecan Street Inc. (2013b) *The Pecan Street Project*. Austin, TX: Pecan Street Inc.

Pecan Street Inc. (2013c) SDG&E Research Study To Dig Down To Circuit-Level Usage. www.pecanstreet.org/2013/08/sdge-research-study-to-dig-down-to-circuit-level-usage/

Pecan Street Inc. (2013d) *What is Pecan Street*. Austin, TX: Pecan Street Inc.

Pecan Street Inc. (2014) Wiki-energy.org Provides University and NGO Researchers Access to the World's Largest Energy and Water Use Database. www.pecanstreet.org/2014/03/wiki-energy-org-provides-university-and-ngo-researchers-access-to-the-worlds-largest-energy-and-water-use-database/

Perera, D. (2012) DoD Aims for Self-reliance with SPIDERS Microgrid. *Fiercegovernment IT*. www.fiercegovernmentit.com/story/dod-aims-self-reliance-spiders-microgrid/2012–07–24

Price, A. (2014) U.S. Energy Secretary Ernest Moniz Visits Austin, With Praise. *Austin American-Statesman*. www.statesman.com/news/news/national-govt-politics/us-energy-secretary-ernest-moniz-visits-austin-wit/ndGHf/

Rifkin, J. (2014) *The Third Industrial Revolution: How Lateral Power is Transforming Energy, The Economy, and the World*. New York: Palgrave Macmillan.

Robertson-Bryan, I. (2011) *Boulder Municipal Utility Feasibility Study*. www-static.bouldercolorado.gov/docs/Feasibility_Study_and_Addendum_81111_net1-1-201305160947.pdf

Sandia National Laboratories. (2012) SPIDERS: The Smart Power Infrastructure Demonstration for Energy Reliability and Security. *Energy, Climate & Infrastructure Security*. energy.sandia.gov/wp/wp-content/gallery/uploads/SPIDERS_Fact_Sheet_2012-1431P.pdf

Schwartz, A. (2009) Energy 101: Austin's Pecan Street Smart Grid Project. *Inhabitat*. www.inhabitat.com/energy-101-the-pecan-street-project/

Seminole Financial Services. (2014) *Renewable Energy Projects*. Bellair Bluffs, FL: Seminole Financial Services. www.seminolefinancialservices.com/projects-financed/renewable-energy-projects/

Sustainable Business. (2013) Community Solar to Double Minnesota Solar Output. *Sustainable Business.com*. www.sustainablebusiness.com/index.cfm/go/news.display/id/25250

Technical University of Denmark. (2014) *Center for Electric Power and Energy.* Lyngby, Denmark: Technical University of Denmark.

The Regulatory Assistance Project. (2011) Electricity Regulation in the US: A Guide. www.raponline.org/docs/RAP_Lazar_ElectricityRegulationInTheUS_Guide_2011_03.pdf.

Trotter, C. (2008) Boulder, CO: America's First Smart Grid Town. *Inhabitat.* inhabitat.com/boulder-co-america%e2%80%99s-first-smart-grid-town/ (last accessed 20 March 2014).

Tweed, K. (2013) Where's My Microgrid? Changes to Energy Markets Could Make Microgrids More Economically Viable. *Green Tech Grid.* www.greentechmedia.com/articles/read/wheres-my-microgrid

U.S. Environmental Protection Agency. (2014) *2013 Climate Leadership Award Winners.* www.epa.gov/climateleadership/awards/2013winners.html#boulder

United Nations Environment Programme. (2012) *Energy.* www.unep.org/climatechange/mitigation/Energy/tabid/104339/Default.aspx

University of Texas. (2014) *Sustainability Directory.* Austin, TX: University of Texas at Austin.

Utilipoint International Inc. (2011) *Critique of Boulder's Feasibility Analysis of Acquiring the Electric Utility Business within the City.* Albuquerque, New Mexico: Utilipoint International, Inc.

Velkomstcenter, B. (2013) *Destination Bornholm.* Ronne, Bornholm, Denmark: Bornholms Welkomstcenter. www.bornholm.info/en

Webster, F. (2001) *Culture and Politics in the Information Age: A New Politics?* London: Routledge.

Windustry. (2014) *Community Wind.* Windustry & Great Plains Windustry Project. www.windustry.org/community-wind

Wright, M. (2012) Are We on the Cusp of a Third Industrial Revolution? *Green Futures Magazine.* www.forumforthefuture.org/greenfutures/articles/are-we-cusp-third-industrial-revolution

Xcel Energy. (2008) *Smart Grid City.* www.xcelenergy.com/About_Us/Energy_News/News_Archive/Xcel_Energy_begins_work_on_SmartGridCity_in_Boulder

Xcel Energy. (2014) *Solar Gardens.* www.xcelenergy.com/Save_Money_&_Energy/Residential/Renewable_Energy_Programs/Solar_Gardens_-_MN

8

A Changing Climate and a Smarter Grid: Critical Linkages[1]

8.1 Responding to Climate Change

Throughout this book we have mentioned climate change multiple times, but we have not explicitly explored it as a motivational force for transforming electricity systems. Several of the most tantalizing promises of smart grid relate to its potential contribution to climate change mitigation, climate change adaptation, or both. Although some smart grid proponents are not concerned about climate change, the creative responses to climate change that smart grid embodies motivates a diverse set of smart grid supporters, ranging from environmental groups such as the Environmental Defense Fund to the U.S. Department of Defense and the European Commission (see previous chapters, especially 6 and 7). Smart grid could facilitate climate change mitigation by enabling a transition away from fossil fuels toward a low-carbon or renewables-based energy system through more dynamic and sophisticated management and monitoring. Smart grid could also facilitate climate change adaptation because the more dynamic and sophisticated approaches to management and monitoring also enable more effective preparation for the increasingly frequent and more intense storms and droughts that threaten electricity system reliability. A smarter grid is viewed by many as a technologically essential part of addressing climate change.

Responses to climate change are often divided into mitigation and adaptation. Climate change mitigation refers to steps taken to reduce greenhouse gas emissions in order to decrease the atmospheric concentrations which are disrupting the Earth's energy balance. Justification of mitigation efforts generally include acceptance of climate scientists' claims that, although climate change is a natural process, human industrial activity and the burning of fossil fuels has intensified and increased the rate of change (IPCC, 2014). The assumption is that, as with ozone depletion (US EPA,

---

[1] This chapter is adapted from and builds on a paper coauthored by the authors of this book as well as one additional collaborator: James Meadowcroft of Carleton University in Ottawa, Canada. See Stephens, J. C., E. J. Wilson, T. R. Peterson, and J. Meadowcroft. (2013) Getting Smart? Climate Change and the Electric Grid. *Challenges* 4, 2: 201–216.

2011), appropriate changes in industrial and social practices can at least partially repair the damage. Climate change adaptation, on the other hand, does not necessarily recognize the importance of anthropogenic contributions to current rates of climate change. Rather it focuses on steps taken to adjust vulnerable infrastructure and social systems to inevitable or already occurring impacts of a changing climate. As of 2014, a strong scientific consensus endorses strategic integration of mitigation and adaptation efforts as the most effective way to approach climate change (IPCC 2014).

Electricity system change, including smart grid, is a crucial component of both climate mitigation and adaptation. A smarter grid has potential to simultaneously contribute to mitigation, by reducing greenhouse gas emissions from the electric sector, and to adaptation, by strengthening the resilience and robustness of electricity systems. By allowing expansion of the proportion of electricity produced by low-carbon renewable resources and enhancing efficiency, smart grid can contribute to climate change mitigation. By strengthening resilience, smart grid can reduce electricity system disruptions during extreme weather events, thus contributing to adaptation.

As we consider the many connections between smart grid and our impact on and response to the Earth's changing climate, it is important to remember that many different smart grid structures and architectures are possible. Some technological configurations of smart grid will clearly contribute to confronting climate change, while other configurations in some places may inadvertently increase climate vulnerabilities. How much a future smart grid electricity system enables climate change mitigation and/or adaptation will depend critically on which actors and public policy priorities shape the design and operation of emerging systems.

This chapter explores how different smart grid configurations may contribute to climate change mitigation and adaptation. We begin by reviewing the continuing societal struggle on how to respond to climate change, including both mitigation and adaptation responses. We then present the ironic possibility that a smarter grid could inadvertently increase, rather than decrease, risks associated with climate change. We then discus two key tensions in linkages between smart grid and climate change: (1) whether smart grid should encourage a more centralized or more decentralized electricity system; (2) whether smart grid should be envisioned as an incremental, evolutionary change or a more radical revolutionary change. We conclude by suggesting strategies for aligning smart grid development with climate change mitigation and adaptation.

8.2 Continuing Societal Struggles

Climate change is a growing threat to the stability of societies throughout the world (IPCC 2014). Although a coordinated international response to climate change remains elusive, national, subnational, and local governments around the world are confronting climate change in different ways, including responding to impacts of climate change and creating policies and incentives for reducing greenhouse gas emissions causing climate change. Despite limited climate policy action in the United States and a strong tendency in U.S. federal politics to avoid the controversial issue of climate change, its realities are becoming more obvious, especially with Superstorm Sandy in October 2012 and other anomalous weather such as the Texas, Midwest, and California droughts of 2013/2014; the "Polar Vortex"; and the ice storms affecting the South and Midwest in winter 2013/2014. The Third National Climate Assessment report released by the U.S. government in May 2014 provides detailed projections of continued trends of warming, heavier precipitation, extreme heat events, more frequent and intense drought, decline in summer Arctic sea ice, and sea-level rise (Walsh et al. 2014). The European Union has taken a more proactive stance on climate change than the United States in terms of both incentivizing reductions in greenhouse gas emissions and improving infrastructure to enhance climate resilience (Biesbroek et al. 2010). Yet major challenges and limited effectiveness of some EU climate policies have resulted in much controversy and continuing societal struggles in Europe on how to address climate change.

8.2.1 Climate Change Mitigation

One of the most prominent smart grid visions includes a massive scaling up of renewable electricity generation to displace much of the CO_2-emitting fossil fuel reliance so embedded within current electricity systems. CO_2 is the dominant greenhouse gas contributing to climate change, and electricity generation is the single largest source of CO_2 emissions. In the United States, the electricity system emits roughly 40 percent of all greenhouse gas emissions (U.S. EPA 2013). The climate change mitigation-focused smart grid vision includes CO_2 emissions reductions from enhancing systemwide efficiency and reducing total generation through storage, grid-side management, and demand-side management. Advanced sensors, as mentioned in Chapter 2 are a key smart grid technology for climate change mitigation. These sensors can facilitate distributed generation with two-way communication throughout the grids, linking local electricity supply and demand response with new demand management tools and smart meters in homes and businesses, and smart household appliances that automatically adjust electricity consumption. This vision often includes more high-voltage transmission lines and energy storage technologies that ease the integration of variable renewable sources by enabling electricity generated at off-peak hours to be shared across broader regions (transmission) or stored for later use. The mitigation potential of smart grid therefore relates in multiple ways to smart meter installation and the better management of energy use (Chapter 5), integration of renewable energy including large-scale wind (Chapter 6), and small-scale community-based grid innovation and microgrids for energy management (Chapter 7).

Climate change mitigation is a powerful motivator for multiple smart grid technologies and actors.

8.2.2 Climate Change Adaptation

As more frequent and intense storms, droughts, and heat waves and other weather anomalies occur with the changing climate, improving the resilience of electricity systems to weather-related disruptions is another powerful motivator for developing smart grid. The notion of climate change preparedness resonates widely as people around the world and across the political spectrum are increasingly aware of the growing threats of and vulnerabilities to climate change. A critical piece of climate change adaptation and preparedness in the United States, at least, involves "system hardening" and resilience by improving the processes and timeframes for recovery from inevitable power outages. Advanced sensors, which detect fluctuations in power flows and identify system irregularities, are a key part of the climate change adaptation component of smart grid because these sensors enable enhanced management, minimize power outages, and help with rapid system recovery.

The devastation and energy system disruptions of Superstorm Sandy demonstrated to the United States, and the rest of the world, the vulnerability of electricity systems. As we mentioned in the very first chapter of this book, the electricity system disruption from this extreme weather event impacted households and businesses across seventeen states, including those as far west as Michigan. The storm left some without power for weeks, and lower Manhattan was in the dark for several days, closing the New York Stock Exchange for two days. Due to the tangible and near-term impacts of electricity system disruptions such as those experienced after Superstorm Sandy, for many people the need for enhanced resilience offers more compelling justification to support smart grid innovation and investments than does the need to reduce CO_2 emissions.

8.3 Ironic Linkages Between Smart Grid and Climate Change

An irony of the complex smart grid system change envisioned by some is that it is possible that smart grid could exacerbate rather than ameliorate the risks of climate change. Although unlikely, it is possible that a future, more sophisticated electricity system could lead to *greater* consumption of high carbon-emission electricity. It is also possible to envision a scenario where an efficient and convenient smart grid system might encourage households and commercial customers to use more electric power. For example, some smart grid configurations could result in the increased use of novel electric gadgets and devices, and the increased use of electricity for transportation, including both personal electric vehicles and public transit. This could result in a net increase in CO_2 emissions depending on how much low-carbon electricity generation is integrated into that particular system. If a smart grid system encouraged increased electrification, the increased use of electricity could also end up negating efficiency gains, an example of the "rebound effect" (Frondel and Vance 2013). The current situation in Germany, described in Chapter 6, exemplifies the ironic potential and unanticipated consequences of multiple energy system and smart grid policies increasing coal use. The German version of smart grid as interpreted and defined within the national-level *Energiewende* involves giving priority to renewable energy and phasing out nuclear, which has inadvertently resulted in increased CO_2 emissions due to an increase in the use of coal – at least in the short term.

It is also possible to envision a future smart grid system that could lock in new vulnerabilities. Smart grid infrastructure and related electricity production/consumption patterns could diminish system robustness and adaptive capacity in the face of an altered climate. For example, the advanced smart grid sensors and internet-based electronic communication could create new system vulnerabilities, with unanticipated negative impacts of these technologies during more extreme weather events, including storms and floods. Efforts to manage electricity demand may also become more challenging because of potentially fluctuating needs (in cold snaps, heatwaves, etc.) and production disruptions (water shortages leading to reduced generation from hydro plants and a lack of cooling water for thermal and nuclear plants). We do not offer these ironic possibilities for the purpose of deterring development and implementation of smart grid, but rather as a cautionary reminder that the technological potential of smart grid could contribute to both climate change mitigation and adaptation efforts in multiple ways. Given the current uncertainty of smart grid development, these scenarios remain a possibility. Their realization requires conscious awareness of the complex interconnections between materiality and symbolicity, between geophysical and sociopolitical dimensions. As Jasanoff and Kim remind us, the sociotechnical imaginaries we invoked in Chapter 2 "are associated with . . . the selection of development priorities, the allocation of funds, the investment in material infrastructures, and the acceptance or suppression of political dissent" (Jasanoff and Kim 2009 p. 123).

8.4 Key Tensions in Smart Grid and Climate Change

The inspiring and optimistic smart grid visions described in Chapter 2 have been invoked widely, particularly by smart grid proponents in industrialized countries, who seek to encourage investment and mobilize action for electricity system change. Yet as the previous chapters in this book illustrate, multiple challenges to advancing smart grid have emerged across jurisdictions and among key societal actors. We see two fundamental tensions that influence relationships between smart grid and climate change, and that provide a framework to map divergent smart grid priorities: (1) whether smart grid should advance a more centralized or a more decentralized

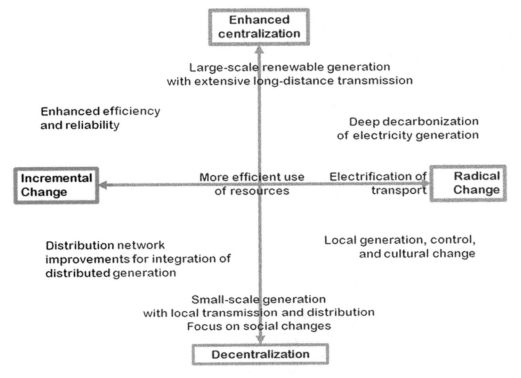

Figure 8.1. Different visions of the potential of Smart Grid can be characterized by perceptions of the possibility and need for enhanced centralization or decentralization and the perceptions of the possibility and need for radical versus incremental change. Source: Stephens et al. 2013

electricity system; (2) whether smart grid should entail incremental, evolutionary change or more radical, revolutionary change (Figure 8.1).

8.4.1 Centralization Versus Decentralization

Whether smart grid is good for the climate depends on many factors, including if smart grid systems are designed to facilitate a more centralized or more decentralized electricity system. Many sustainability advocates would argue that decentralization and localization of all systems, whether food systems or energy systems, should be a guiding principle for both climate change mitigation and climate change adaptation (Lovins 1977; Rifkin 2011). But this principal conflicts with others who call for the need to consider scaling up, intensification, and efficiency in all systems to effectively meet the needs of all the people on the planet. Those who share this perspective tend to view smart grid as a key piece of enhanced centralization and system function involving larger interconnected electricity systems that rely on expanded long-distance transmission and distant large-scale electricity generation far from demand centers.

An increasingly centralized electricity system may enable the development of large-scale renewable generation where the resources are available through the use of efficient long-distance high-voltage transmission lines to move the power hundreds of miles to sites where the electricity is needed (see Chapter 6). These networked systems will link multiple private and public sector actors to develop, manage, and maintain the systems. For example, the proposed (and now abandoned) DesertTec project anticipated powering much of Europe with electricity generated from concentrated solar power located in North Africa (Desertec 2014). Some North American proposals envision an extensive high-voltage transmission grid overlay to move large-scale and inexpensive Midwestern U.S. wind to energy markets in the more populated coastal regions. Such systems would allow for economies of scale and development of areas with strong renewable resources. They could also benefit some incumbent energy sector actors, though regional differences and context-specific factors would determine which actors gain or lose. For example, East Coast utilities could see their fleet investments undercut by cheap Midwestern wind-generated electricity, but East Coast consumers could see electricity prices decrease (Piller 2010).

An increasingly decentralized electricity system is a priority for other societal actors who support more local distributed generation and community control to encourage electricity production and economic development close to demand centers (Verbong and Geels 2010; Wolsink 2012). The Danish Energy Association, for example, includes smart grid in its goals for national energy independence, replacement of fossil fuels, and integration of massive amounts of renewable energy, often generated and distributed at residential or municipal levels (Pentland 2008; Ullegård 2013). Decentralization, including microgrids and local renewable production such as rooftop solar PV, is heralded as providing customer and community empowerment and potentially lessening centralized corporate control of electricity systems.

Investment focused on reorienting the grid toward either of these endpoints reduces the likelihood of achieving the other: if investments are made in local electricity generation, the demand for long-distance transmission lines and centralized generation will be reduced. On the other hand, major investment in long-distance transmission lines and centralized electricity generation at sites far from demand centers could reduce the need for distributed local generation. As discussed in Chapter 7, decentralization also collides with existing patterns of ownership and control and, given the power and expertise embedded in established institutions that rely on a centralized system (Munson 2005), a widespread shift to decentralization may be difficult (Wolsink 2012; EEI 2013). The controversy in Boulder, Colorado (Chapter 7) highlights this tension: Xcel Energy, the investor-owned utility that serves Boulder, initiated a smart grid demonstration project in response to community demands for more low-carbon sourced electricity, but cost overruns and the utility's continued reliance on coal – even as Xcel has incorporated high levels of renewables into its generation mix – has frustrated residents. The city is currently exploring municipalizing its electricity services as a more effective means of meeting its goals for climate change mitigation.

From a climate change perspective, strategic combinations of centralization and decentralization appear likely to contribute to deep GHG emission reductions for climate change mitigation and/or to enhanced resilience for climate change adaptation. In some places, decentralization could empower communities to create resilient linked distributed generation and demand management and move to a lower-carbon and/or less vulnerable local system, while other regions may embrace large-scale generation of low-carbon energy distributed and managed through a more centralized system. But, of course, actors have reasons for favoring centralization or decentralization besides concern with climate change. The heterogeneity in the geographic scale and scope of electricity system development means that actors typically approach smart grid priorities based on what appears optimal from a narrow jurisdictional context, with few considering the implications for larger or smaller physical/administrative scales. How the sociopolitical context, and its business models, laws, regulations, and policies, promote different actor interests in creating a smart grid links to the system design and outcomes.

8.4.2 Incremental, Evolutionary Change Versus Transformative, Revolutionary Change

In addition to the struggle over whether smart grid should promote more centralization or more decentralization, a second tension relates to the extent to which smart grid implies incremental improvements to the existing system – evolution – or dramatic system transformation – revolution. Smart grid is viewed by many as a gradual process of energy system modernization, geared to optimize current ways of providing electricity. Yet many others have argued that responses to climate change require transformative rather than incremental changes (Rifkin 2014). And some key actors view smart grid as a radical shake-up that includes novel technologies, new operating procedures, and the establishment of new norms, expectations, and business models.

Just as the centralization and decentralization dichotomy may require both, smart grid may need to simultaneously offer a radical, long-term vision of change and more immediately *practical operational* changes that represent a steady and incremental "smartening" of existing systems. Perhaps both are possible and necessary. While some actors – especially those interested in climate goals – emphasize the disruptive potential of smart grid technologies to dramatically transform the way we make and use electricity, others who are more involved in system operation emphasize smart grid as a series of incremental step-changes to address multiple energy-related problems.

Our research has shown that established electricity system actors (especially utilities and associated regulators) are more likely to define smart grid in terms of incremental rather than radical change. An extensive set of focus groups with multiple electricity sector stakeholders in multiple states and regions in the United States revealed that those whose jobs and organizations focus on the day-to-day operations of electricity systems have a more incremental perspective on the potential of smart grid, while those with a broader societal view, including environmental groups and energy system researchers, tend to focus more on the long-term transformative potential of smart grid to change the energy system. Those involved in day-to-day system operations tend to be suspicious of grandiose schemes with uncertain risks and benefits and are often wary of upsetting customers with increased bills. These incumbent actors also have the most at risk from rapid innovation associated with the entry of new actors into the electricity system. Moreover, electricity system engineers focused on maintaining day-to-day operations are cautious about innovations that might compromise system reliability.

Environmental and climate advocates and energy researchers, on the other hand, often focus on long-term issues and tend to emphasize the potential for radical change without always anticipating the social and institutional obstacles to systemic socio-technological change. One example of this can be seen in the practical challenges associated with implementation of the Cape Wind project. Although the development of offshore wind in Nantucket Sound offers a climate-friendly way to provide carbon-free electricity to eastern Massachusetts, the scale of the long, expensive, and juris-dictionally complex controversy that has slowed down the Cape Wind project was not widely anticipated within energy and climate communities. These tensions are typical of sociotechnical transitions wherein change is resisted by multiple actors for an array of reasons. Unlike Bornholm, Denmark, another picturesque tourist island where wind development was widely supported by local residents, the political controversy over Cape Wind damaging wealthy resident viewsheds does pose a certain environmental irony. Senator Edward Kennedy – a longtime environmental advocate and crusader against pollution from Appalachian coal – was an active opponent of Cape Wind.

8.5 Linking Smart Grid and Climate Change

The breadth of different priorities among societal actors supportive of smart grid innovation allows the potential for synergistic alignment of interests including climate change mitigation, climate change adaptation, and other societal objectives. This broad spectrum of priorities also highlights the risk of climate priorities being neglected in smart grid development. The question of how distributed smart grid developments in specific contexts can be synergistic with climate change objectives depends on which key actors have the most influence on smart grid development and how smart grid architecture is structured.

Specific strategies to ensure climate priorities are integrated into smart grid deployment must be tailored to fit region-specific contexts. Coal-heavy systems such as those in the U.S. Midwest, Poland, or the Canadian province of Alberta present different challenges and opportunities for smart grid than hydro-dominated systems such as those in Norway, the Canadian province of Quebec, or the U.S. Pacific Northwest. Restructured electricity markets, traditionally regulated systems, and government-owned power companies each present different business opportunities and logics for smart grid development. And local or national political constraints or resource endowments may favor particular sets of smart grid configurations. The context of innovation for electricity systems is critically important; leverage points which could link smart grid and climate change in one set of circumstances could have the opposite effect in another and unintentionally subvert climate objectives.

Accepting the importance of local and regional contexts, our analysis of the connections between smart grid and climate change have encouraged us to develop six principles that we think could be applied to smart grid priorities in any context to encourage smart grid development that is synergistic with climate change objectives (Stephens et al. 2013).

(1) At the planning stage, we recommend that all smart grid investments be formally assessed for potential contributions to climate change mitigation and adaptation in both the short and long term. This accounting for the climate implications of electricity system investments could be a government requirement integrated into financing and regulation to guide a long-term trajectory of smart grid rollout that places explicit value on both climate change mitigation and adaptation.

(2) Smart grid initiatives that contribute to energy efficiency and electricity conservation should be prioritized because controlling demand is often the cheapest and most effective way to reduce both GHG emissions and costs.

(3) Smart grid initiatives that facilitate the incorporation of low-carbon generation should be prioritized.

(4) Smart grid measures that support the emergence of local microgrids and enhance local and community-based energy systems are generally positive, but should also be evaluated in concert with local air pollution goals and energy system reilability. Bringing generation close to the point of use reduces transmission losses and allows the development of integrated energy solutions (multiple fuels, combined heat and power, etc.) in buildings and local communities. As long as it does not compromise local air quality, localization could also allow for more modular and, therefore, more adaptable systems.

(5) When it comes to smart grid operation, particular attention should be paid to ways it can enhance system flexibility and redundancy. Climate uncertainty and the unpredictability of future energy needs point to the importance of adaptive management approaches (that can make rapid adjustments in response to fuel price changes, resource shortages, or technical disruptions) – smart grid innovations can be helpful here.

(6) Smart grid proponents need to promote a culture of transparency. This means presenting and evaluating the specific economic, environmental, and social benefits particular smart grid investments will secure. Ideally, proponents should seek to avoid cycles of hyperbole and subsequent disappointment. They need to shun schemes which support monopolization of smart grid gains or benefits to particular interests, while socializing the costs. This relates to the questions about who controls and who benefits from smart grid that we have addressed throughout the book.

Smart grid is a critical part of a widespread societal push for an energy transition that is unlike past energy system transitions because of the dual motivations of climate change mitigation and adaptation. This push toward an energy transition is motivated by sustainability and climate change mitigation on the one hand and resilience and climate change adaptation on the other (Hess 2013).

While these multiple connections between smart grid and climate change offer some possibilities for developing a unifying smart grid vision across different actors, the contentious nature of climate change continues to highlight fundamental societal tensions.

8.6 Conclusions

The guiding question of this chapter asked what configurations of smart grid are most valuable to climate change mitigation and adaptation. To answer that question we have explored some of the complex relationships among smart grid, climate change mitigation, and climate change adaptation. Both the local and regional contexts influence assessments of the most valuable ways to respond to climate change. A smart grid improvement in one community may help that community respond to climate change; that same improvement may make it more difficult for another community, state, region, or country to respond to climate change. At a global scale some argue that all efforts to promote renewable electricity generation are good for climate change, but within this book we have demonstrated flaws in that simple assertion.

Beyond climate change mitigation and adaptation, many other issues motivate interest in smart grid. Throughout the previous chapters of this book we explored many of these other motivating factors, including reduced electricity costs, improved efficiency, increased electricity access, minimized electricity theft, and enhanced energy security. Other environmental benefits beyond climate change are also motivating factors for smart grid, including reducing negative environmental impacts of coal, gas, and oil extraction and combustion on human health and biodiversity. Ensuring that smart grid will help society respond to climate change requires creative synergistic integration of climate objectives with other economic, social, and environmental objectives.

The complexity of smart grid illustrates both the challenges and the opportunities of integrating climate change priorities into broader societal and user goals. The smart grid story highlights how important it is for those advocating for climate policy to connect their priorities with other societal objectives. Across multiple venues, climate priorities are not part of the dominant decision making logic of many electricity system actors; rather, they must be explicitly woven into initiatives and policies that are simultaneously advancing other economic, social, and environmental objectives. The vague inclusiveness of the term smart grid appeals to a diversity of supporters, but also obscures actors' divergent values and system endpoints. General smart grid debates often omit the critical details of how future electricity systems will develop;

yet these details remain essential if smart grid is to help societies confront and adapt to climate change.

Investment in a smarter grid is happening now. Responding and adapting to climate change is a critical societal imperative, and smart grid design must reflect and integrate climate goals into its design and operation. Until climate goals are explicitly embedded within formal electricity system decision making structures, smart grid development may perpetuate growing greenhouse gas emissions by strengthening the dominant energy system growth paradigm of increasing electricity generation and use. Getting smart about linking electricity system change and climate change objectives is an urgent societal priority.

So, how to negotiate the tensions we have highlighted in this chapter in ways that nudge smart grid toward enabling both climate change mitigation and adaptation? While climate objectives can be integrated into both centralized and decentralized systems, climate goals cannot ultimately be achieved without radical changes in the ways electric power is produced and consumed. Given the scale of the climate problem, social and institutional as well as technical changes in energy systems will be required. When considering such transformative change, a fundamental challenge is the extent to which incremental improvement and more radical system transformation can be reconciled. While incremental adjustments can bring immediate gains, and contribute to broader patterns of system change, in certain circumstances they can also defer more radical innovation, and even enhance lock-in to a sub-optimal development trajectory. In large, complex, and interconnected systems like the electric power sector, poorly conceived incremental changes can work against long-term goals. For smart grid to be effectively linked to climate change objectives, short term implementation priorities must be established with a clear eye on the long-term, more fundamental goals of transforming electricity systems to be low-carbon and resilient.

References

Biesbroek, G. R., R. J. Swart, T. R. Carter, C. Cowan, T. Henrichs, H. Mela, M. D. Morecroft, and D. Rey. (2010) Europe Adapts to Climate Change: Comparing National Adaptation Strategies. *Global Environmental Change*, 20, 440–450.

Desertec. (2014) Desertec Foundation. Desertec Foundation. Heidelberg, Germany, www.desertec.org/

Frondel, M. and C. Vance. (2013) Energy Efficiency: Don't Belittle the Rebound Effect. *Nature*, 494. www.nature.com/nature/journal/v494/n7438/pdf/494430c.pdf

Hess, D. J. (2013) Transitions in Energy Systems: The Mitigation-Adaptation Relationship. *Science as Culture*, 22, 144–150.

IPCC. (2014) Climate Change 2014 Mitigation of Climate Change. Intergovernmental Panel on Climate Change. www.ipcc.ch/report/ar5/wg3/.

Jasanoff, S. and S. H. Kim. (2009) Containing the atom: Sociotechnical imaginaries and nuclear power in the United States and South Korea. *Minerva*, 47(2), 119–146.

Kind, P. (2013) Disruptive Challenges: Financial Implications and Strategic Responses to a Changing Retail Electric Business. Washington, DC: Edison Electric Institute. www.eei.org/ourissues/finance/documents/disruptivechallenges.pdf

Lovins, A. B. (1977) *Soft Energy Paths: Towards a Durable Peace*. New York: Harper Colphon.

Munson, R. (2005) *From Edison to Enron: The Business of Power and What it Means for the Future of Electricity*. Westport, CT and London: Praeger.

Pentland, W. (2008, August 7) Distributed Energy: The Answer to the Energy Problem. *Forbes*. www.forbes.com/2008/08/06/denmark-energy-electricity-biz-energy-cx_wp_0807power.html

Piller, D. (2010) Eastern Governors Protest Midwest Wind Transmission Line. *DesMoines Register*. blogs.desmoinesregister.com/dmr/index.php/2010/07/13/eastern-governors-protest-midwest-wind-transmission-line/article

Rifkin, J. (2011) *The Third Industrial Revolution: How Lateral Power Is Transforming Energy, the Economy, and the World*. New York: Palgrave Macmillan.

Stephens, J. C., E. J. Wilson, T. R. Peterson, and J. Meadowcroft. (2013) Getting Smart? Climate Change and the Electric Grid. *Challenges*, 4, 201–216.

U.S. EPA. (2013) *Global Greenhouse Gas Emissions*. Washington D.C.: U.S. EPA. www.epa.gov/climatechange/science/indicators/ghg/global-ghg-emissions.html

U.S. EPA. (2011, 26 April). The Science of Ozone Layer Depletion. www.epa.gov/ozone/science/

Ullegård, J. (2012, July 11) Commission Reiterates Focus on Smart Grid. Copenhagen, Denmark: Danish Energy Association. www.danishenergyassociation.com/Theme/Smart_Grid.aspx

Verbong, G. P. J. and F. W. Geels. (2010) Exploring Sustainability Transitions in the Electricity Sector with Socio-technical Pathways. *Technological Forecasting and Social Change*, 77, 1214–1221.

Walsh, J., D. Wuebbles, K. Hayhoe, J. Kossin, K. Kunkel, G. Stephens, P. Thorne, R. Vose, M. Wehner, J. Willis, D. Anderson, S. Doney, R. Feely, P. Hennon, V. Kharin, T. Knutson, F. Landerer, T. Lenton, J. Kennedy, and R. Somerville. (2014) Our Changing Climate. Climate Change Impacts in the United States. In *The Third National Climate Assessment*, eds. J. M. Melillo, T. T. C. Richmond, and G. W. Yohe. U.S. Global Change Research Program, 19–67. nca2014.globalchange.gov/report

Wolsink, M. (2012) The Research Agenda on Social Acceptance of Distributed Generation in Smart Grids: Renewable as Common Pool Resources. *Renewable & Sustainable Energy Reviews*, 16, 822–835.

9

Smart Grid (R)evolution

9.1 Inevitable but Unpredictable Change

We are experiencing inevitable, ongoing major changes in our electricity systems. Whether one views the potential of smart grid as revolutionary, evolutionary, both, or neither, energy system change is happening. Smart grid will continue to expand, develop, and evolve as individuals, communities, utilities, states, regions, and countries struggle to integrate multiple new and emerging challenges and expectations for energy and electricity systems. This electricity system transition is being influenced by a confluence of forces that is simultaneously encouraging change in technologies, institutions, and culture. Who has control and who benefits from these changes is also shifting; how, when, and where control is exerted and benefits are realized is a dynamic, context-specific evolution with revolutionary potential.

We began this book in Chapter 1 reflecting on tensions, struggles, and opportunities associated with the electricity system disruption along the northeastern seaboard of the United States during and after Superstorm Sandy in October 2012. The impact of that storm reached beyond the observable economic and physical damage to highlight to the world societal vulnerabilities in an era of increased electricity dependence and increasingly vulnerable infrastructure facing more frequent and intense storms. Among the multiple tensions that emerged in the aftermath of the storm, the struggle to figure out how best to restore, protect, and transform the electricity system to minimize the likelihood of a similar future disruption has been one of the most challenging.

We optimistically suggested in Chapter 1 that understanding how power struggles are developing and why tensions are evolving can contribute to creative alignment of interests and priorities in different places. As we now conclude this book, we maintain this optimism with a call for broader, more inclusive, and more imaginative conversations about smart grid and its social implications. This final chapter justifies expanding smart grid conversations and collaborations by first highlighting the importance of a sociotechnical perspective in considering smart grid. We focus on the power struggles associated with who has control and who benefits from smart grid, followed by a Canadian example of evolutionary smart grid innovation paving the way for revolutionary change. Finally, we acknowledge the diversity of smart grid futures and conclude with a simple and practical message: broadening smart grid conversations will advance collaborative thinking and engagement on the social implications of electricity system change.

9.2 Encouraging a Sociotechnical Systems Perspective on Smart Grid

Power outages of any kind, whether they are multiple-day blackouts like the disruptions following Superstorm Sandy or shorter outages of just a few hours, remind us how the flow of electricity influences our communication, our culture, and our communities. Despite general awareness of the social implications of energy, many still think of electricity system change in mostly technical and economic terms. Smart grid, with all its social complexity, provides an opportunity to refocus and expand beyond this narrow technical–economic lens. When we expand this lens we see that some strands of smart grid conversations echo Amory Lovins' 1977 notion of soft energy paths that integrate evolving and malleable societal needs into energy planning to encourage systems that are flexible, responsive, benign, and sustainable (Lovins 1977). With a broader lens we also see that smart grid offers a path toward Jeremy Bentham's Panopticon, where energy use data can become another tool for constant surveillance tracking our activities and movements, and mapping our desires (Bentham 1995). Our review of the multiple promises and pitfalls of smart grid in Chapter 2 highlights the diversity of perspectives and perceived social implications of smart grid. Both positive and negative possibilities of smart grid can be envisioned when smart grid is considered as part of a larger and ever-changing sociotechnical system.

The tendency to view electricity systems through a purely technical–economic lens often obfuscates the larger societal dimensions of these systems. Too often consideration of electricity system change is delegated to engineers and economists, who work within a narrow focus and are trained to limit their analysis to physical practicalities and economic costs of technical systems. With this book we have attempted to widen this focus to incorporate nontechnical conversations about the social dimensions of smart grid.

Throughout the book we have acknowledged that smart grid embodies different kinds of changes for different people; it is a term with multiple simultaneous meanings. For some, smart grid signifies a technological nirvana, a bucolic end-state where happy people will drive their electric cars past wind turbines while taking deep breaths of clean air. For others it is an empty signifier, a term so broad and vague as to be meaningless. For some, it means incremental improvements to maintain the status quo, yet for others it has become part of a revolutionary social movement; part of a larger effort to redistribute power and control and counter growing disparities between rich and poor. Just as there are different kinds of human intelligence (social intelligence, mathematical intelligence, emotional intelligence), a smart grid can be considered "smart" in many different ways and for many different purposes.

To appreciate the critical importance of the social dimensions of smart grid, it is useful to consider the interconnectedness between smart grid innovation and other technology innovations, particularly in the area of communications. The past decade of innovation in communication technology has been revolutionary. The explosion of smart phone usage and expectations of constant connectivity have revolutionized cultural expectations regarding data, information, and communication access and availability. This is linked directly to our cultural expectations of electricity access. While it is clear to many of us that the internet has democratized information, knowledge, and content creation by providing individuals and communities with access to and control of information, we are only beginning to imagine how a similar trend in distributed electricity generation could democratize energy by giving individuals and communities direct control of energy (Rifkin 2011). Some argue that the internet has also centralized control with information about on-line purchases, Google searches, and physical movements tracked, stored, and subject to search. Both smart grid and the internet revolution could contribute to moving society away from a hierarchical model toward a more lateral structure, incorporating more distributed power and influence – or they could serve to further consolidate access and power in the hands of a few.

9.3 Power Struggles: Who has Control and Who Benefits

Many of the tensions and controversies surrounding smart grid can be attributed to struggles over who has control and who benefits from smart grid improvements. The possibilities for sociotechnical change associated with smart grid offer positive opportunities for some societal actors and negative, threatening future scenarios for others. Among the many societal actors involved in smart grid development that we reviewed in Chapter 4, established actors like the large utility companies have very different interests and priorities than do, for example, self-declared locavolts who are seeking novel ways to generate electricity for themselves. These different actors are striving for different outcomes, different types of control, and different kinds of benefits. One clear distinction is between established, incumbent actors and new actors. Our interviews with a diverse cross-section of agents confirm that incumbents who are well established with years of working within and benefiting from the current electricity system are generally more cautious and conservative about the potential for smart grid change than are renewable energy entrepreneurs or environmental advocates who are eager for transformative change and see smart grid innovation as an important tool. Many incumbent actors are more likely to favor slow and incremental change, while newer actors with less firmly established relationships with existing systems (and less of a financial stake in maintaining the status quo) are more likely to envision radical, even revolutionary, change.

Power struggles surrounding who has control and who benefits from smart grid emerged prominently in our analysis of three particularly important aspects of smart grid development: smart meter deployment in Chapter 5, integration of large-scale wind power in Chapter 6, and community-based and small-scale initiatives in Chapter 7. Smart meters could offer new kinds of control to both electricity consumers and the utilities managing the flow of electricity. Cost and efficiency benefits are possible for both consumers and utilities through the enhanced capacity to monitor and measure electricity use. Some stakeholders are skeptical, however, about whether the economic benefits of smart meters are greater for the utilities or for the consumers. The strong opposition to smart meters that is felt deeply by some reflects concern about a loss of control associated with a mistrust of government's and utilities' concern to protect privacy and health.

The coevolution of smart grid development and the integration of large-scale wind energy explored in Chapter 6 reflect different kinds of power struggles. As we noted, the rapid scaling up of wind power during the last decade in Texas, the Upper Midwest, and Germany has resulted in levels of wind integration previously thought to be technically impossible. But this new renewable generation has also generated multiple struggles over the control of infrastructure development. A recurring challenge in all three of these cases discussed in Chapter 6 focused on who has control over building (and paying for) new transmission capacity to move the power to where it is needed. With transmission lines, additional power struggles emerge because those who benefit from the additional electricity transported through the new power lines are often not the communities impacted by or paying for the new transmission lines.

The community-based and small-scale initiatives explored in Chapter 7 highlight another set of power struggles associated primarily with a quest for more local control and self-reliance. In these examples, we see how unique and strong individuals and communities are working to counter the prevailing paradigm of centralized, large-scale electricity generation. These examples provide inspiring examples of how individuals and communities have been taking control of electricity systems and ensuring that they benefit from the systems in the ways they desire. Boulder, Colorado; Austin, Texas; and Bornholm, Denmark each have a history of engaging in opportunities to develop new approaches to electricity systems. While each of these examples provides a unique story, the quest for more local control in electricity system is a commonality that is emerging in communities in many parts of the world.

A different kind of struggle over the imagined future benefits of smart grid was illustrated in our Chapter 8 discussion of climate change and the environmental uncertainties of smart grid innovation. While environmental improvements are often touted as a dominant promise of smart grid justifying grid innovation, the degree to which smart grid may actually contribute to environmental improvement is neither self-evident nor predetermined. The environmental impacts of smart grid development depend on the details of implementation, which actors' perspectives are most influential, which smart grid configurations are advanced, and what environmental impacts are prioritized. All of these details are context-specific, dynamic, and dependent on which societal actors have more control in shaping smart grid systems. It is possible to imagine, for example, a scenario in which a large utility, heavily dependent on coal-fired power generation, uses smart grid technologies, including syncophasors and new high-voltage transmission lines, to operate their fossil fuel plants at lower cost. Such a scenario exemplifies the possibility that some smart grid futures may not maximize environmental benefits.

9.4 Evolution to Revolution: Wind-to-Heat in a Small Canadian Community

To illustrate the potential for evolutionary smart grid innovation leading to more transformative revolutionary change, we introduce one final example of electricity system change in the Canadian city of Summerside on Prince Edward Island (Belanger 2014). This small city of about 7,000 households and 14,000 people installed a four-turbine 12 MW wind farm in 2009, which was largely a financially motivated decision by the city. Between 2010 and 2012 the city profited by $1.6 million annually from selling the electricity to the local municipal utility, Summerside Electric. Given the variability of the wind, however, the utility ended up having more electricity than it could use locally, and due to limits on its power purchasing agreements it lost money if the excess power was sold elsewhere. To help manage the mismatch between supply and demand and expand the city's use of the local power, the municipal utility worked with the city to explore options for storing and using surplus electricity for residential and commercial heating in Summerside.

Electric thermal storage (ETS), the storage of electricity as heat in an insulated brick core, became the focus of Summerside Electric's strategy, because the excess renewable electricity during off-peak hours could be used to provide a reliable and low-cost steady stream of heat for water and space heating for residential and commercial customers (Belanger 2014). This so-called "Heat for Less" program also required the installation of a fiber network to enable reliable two-way communication and the smart meters necessary to monitor and manage the electricity storage capacity of the ETS units. Although the fiber network involved a substantial city investment, this part of the project gained community support by emphasizing the co-benefit of providing the entire community with a rapid internet connection.

Through a series of evolving decisions, the energy system in Summerside has changed dramatically in just a few years. Citizens' use of heating oil has plummeted as the city's wind capacity now provides about half of the residents' power for both electricity and much of the community's heating needs. This story exemplifies both the evolutionary and unpredictable nature of smart grid innovation and the resulting revolutionary changes in the community's energy system. Investments in one set of technologies (wind power) led to new opportunities and new justifications for supporting other kinds of investments, including communication and social investments that have allowed this community to become more self-reliant in its energy systems as they generate more of their electricity and provide for their heating needs.

The wind-to-heat project in Summerside also highlights the critical role that electricity storage can play in energy systems. Storage could be one of the biggest game-changers in smart grid innovation. When electricity can be stored efficiently at low cost, the system's flexibility increases, in this case opening up new opportunities to connect electricity production and heating needs. Just as the advent of widespread refrigeration changed our cultural expectations, practices, and technologies associated with food and cooking, the advent of widespread electricity storage has potential to change our cultural expectations, practices, and technologies associated with electricity use.

9.5 Diversity of Smart Grid Futures

The Summerside example also showcases how the local context shapes smart grid and electricity system change. There is no cookie-cutter, "one-size-fits-all" smart grid configuration that should be replicated in communities throughout the world. Every place has its own set of smart grid priorities and will have different opportunities and face particular challenges. Throughout this book's exploration of the many dimensions of smart grid, the shift toward diversification of technologies, institutional structures, perspectives, and actors to address specific system challenges has emerged as a central theme.

This diversity is apparent in almost every aspect of smart grid, from the promises and pitfalls (Chapter 2) to the technological components considered central to smart grid (Chapter 3). There is also diversity in priorities and perspectives among societal actors (Chapter 4), as well as diversity within societal actor subgroups. For example, we see diversity among the established large utilities in terms of their orientation and engagement toward smart grid technologies and how they are responding and adapting to new social expectations.

In addition to increased diversity in sources and scale of electricity generation, there is increased diversity in ownership of assets. New opportunities are emerging for individuals, businesses, and communities to participate differently in innovative electricity system changes. Widespread deployment of distributed generation technologies offers enormous potential to change asset ownership models. For example, solar companies such as SolarCity are leasing solar panels on rooftops while maintaining ownership and risk, facilitated by dropping solar PV prices and generous policy incentives. However, not all new energy technologies make immediate economic sense or even produce much electricity. The small-scale wind turbines being installed on high-rise buildings in New York City are appealing to some, and while they may generate enough power to light the building's hallways and lobby, they are viewed by others as largely ornamental signifiers of "green values" designed primarily to make the buildings attractive to renters (Chaban 2014).

Although we acknowledge this diversity of smart grid and its potential futures, we are not embracing an "all-of-the-above" smart grid strategy in the same way that the United States has officially embraced an all-of-the-above energy strategy (Moniz 2013). It is clear that generic support for smart grid could be used to justify almost any kind of proposed electricity system change. Given this diversity of smart grid futures and the interests of societal actors involved in shaping electricity systems, we believe that stakeholders share responsibilities to assess the operational, societal, and environmental consequences of smart grid innovation.

With the hope of assisting smart grid actors in fulfilling this responsibility and broadening their understanding of systemwide change, we have tried in this book to synthesize multiple perspectives without privileging the perspectives or priorities of one set of actors over another. In our research and in our writing of this book, we were not smart grid advocates. We intentionally tried to maintain simultaneous skepticism and enthusiasm about smart grid promises, while we were also cautious and concerned about the potential pitfalls of smart grid. From all that we have learned over the past six years of research, we do not have, or even agree upon, a singular vision of a smart grid future.

Although we recognize and embrace the diversity in smart grid futures, this book focuses on North America and Europe. By concentrating on northern hemisphere, relatively wealthy countries with long-established electricity system infrastructure we were able to delve deeply into specific and somewhat comparable examples. We realize that with this focus we have left out some of the exciting smart grid advances occurring in the rest of the world. Smart grid innovation is taking place for many reasons in many different locations, and the diversity of its many objectives, implementation strategies, and societal actors is rich and expanding.

9.6 Broadening Smart Grid Conversations and Collaborations

Given the diversity of smart grid futures, we do not consider ourselves advocates of a particular smart grid pathway. We are, however, strong advocates of broadening societal conversations about smart grid to enable more diverse participation and collaboration. We believe that when smart grid conversations are expanded beyond the dominant technical and economic perspectives to integrate social and cultural dimensions, a more inclusive set of collective energy system goals can be established to guide smart grid development (Stephens et al 2014). Broader conversations about smart grid encourage a re-examination of our perspectives on what is possible, what is desirable, and why energy systems have developed the way they have. Rather than embrace a naïve, technically optimistic perspective of smart grid potential, we encourage broad societal engagement to address critical infrastructural challenges because how we tackle these issues and who is involved in these conversations will determine the future distribution of power in society (both electrical power and other forms).

We know that all changes, whether technical advances, policy innovations, or cultural shifts, result in second and third-order changes that are impossible to predict or anticipate. There will always be risks associated with both the deployment of new technology and the lack of deployment of new technology. All change is challenging, and implementation of system change requires imagination, flexibility, and adaptation.

One important lesson we draw from the experiences of widespread smart meter deployments across the world, wind and solar deployment in Germany, or efforts to increase community control in Boulder, Colorado, is that change that may have seemed impossible to many can become possible in unexpected ways. When engineers, regulators, or planners declare that something is impossible, it may one day become possible with shifts in technology, as a result of policy incentives, or because someone tried to do it. Engineers, regulators, and planners ten years ago claimed that it would be impossible to have more than 20 percent wind on any electric system because the system could not manage the resource variability and might collapse. In the United States, several states have moved beyond the 20 percent mark, with the state of Iowa leading with 27.4 percent of the state's electricity generated from wind (AWEA 2014). In the EU, similarly high penetration of wind power has been reached in several parts of Denmark and Germany.

This expansion of possibilities requires the simultaneous consideration of multiple political, economic, technical, social, and cultural perspectives (Stephens et al 2008). One of our goals in this book has been to bring these multiple perspectives into conversation by juxtaposing them against one another. Multiple different interactions among societal actors are required for electricity system change, and new opportunities for collaborative engagement to define these changes are emerging. We recognize that not everyone has the interest or time to participate and engage in smart grid development. We know, however, that our communities and the organizations where we work will increasingly be required to invest time and other resources in energy system change, and greater levels of awareness, knowledge, engagement, and participation support deliberate decision making and informed design.

When considering any complex system change such as smart grid, it is clear that no individual or organization, regardless of access to intensive data sets or sophisticated models, can reliably predict specific future outcomes, especially second or third-order changes. In today's rapidly changing, complex, and interconnected world, building capacity to adapt and respond to inevitable future changes is becoming the most valuable asset. Based on our analysis, these are the attributes of smart grid – enhanced capacities for flexibility and resilience while maintaining system reliability – that are the most intriguing and important.

The practical message of this book is that broadening smart grid discussions will advance collaborative thinking on the social implications of electricity system change. As electricity plays an increasingly critical role in our lives, incorporating the social dimensions of energy system transitions into system design and implementation has become a crucial step to addressing societal needs.

One important part of broadening conversations about energy system change involves expanding energy education beyond engineering and technical perspectives (Brummitt et al. 2013). As educators, we are very aware of how educational conventions and disciplinary separation restrict learning about energy. In many places, energy-related courses are limited to the engineering curriculum, and social and cultural impacts of energy system change are not integrated. But every individual, every community, and every organization is impacted by energy system change, so an expansion of energy education could enhance engagement and broaden energy conversations.

Current electricity system practices and strategies have been codified in our institutions and laws, but these practices are not necessarily the strategies that will enable us to most effectively create the energy systems we need for the future. We are in the midst of a systemwide need to reprioritize our strategies and our approaches to planning and operating energy systems. The concurrent demands of climate change mitigation and adaptation, new technological capabilities, and shifting societal needs mean that the strategies, technologies, and expectations of utilities, regulators, and customers will also evolve. The tools and strategies that have been used to ensure that the electricity system is reliable and affordable in past decades are unlikely to be the same as those required to meet new expectations of electricity systems in the coming decades. As we look forward, it seems clear that closely interconnected social and technical changes in electricity systems will continue to develop in complicated and uncertain ways.

We conclude by returning to the Indian parable of the elephant and the blind men that we introduced in Chapter 1. This story represents the limits of any individual's subjective experience in seeing the whole truth, or the whole system, or the whole elephant. Each blind man was only able to apply his limited experience touching one part of the elephant's body (either the tusk, the trunk, the tail, the legs, or the underbelly) to extrapolate and envision the entire animal. We suggest that this same principle holds true with regard to how societal actors are currently engaged with smart grid development. Actors view smart grid from their unique perspective and use this as a base from which to extrapolate and envision a particular smart grid pathway and the future of the electricity system. Given the sociotechnical complexity and dynamic context of smart grid futures, no individual can see the whole system and its potential. Only by appreciating the multiplicity of smart grid perspectives can we successfully engage with the many dimensions of electricity system change in ways that enable us to collaboratively move toward a more positive and sustainable future.

References

AWEA. (2014) Iowa Wind Energy. American Wind Energy Association. www.awea.org/Resources/state.aspx?ItemNumber=5224

Belanger, N. (2014) *A Canadian Smart Grid in Transition: A Case Study of Heat for Less.* Waterloo, Canada: University of Waterloo.

Bentham, J. (1995) *The Panopticon Writings*. London: Verso.

Brummitt, C. D., P. D. H. Hines, I. Dobson, C. Moore, and R. M. D'Souza. (2013) Transdisciplinary Electric Power Grid Science. *Proceedings of the National Academy of Sciences*, 110(30), 12159. www.pnas.org/content/110/30/12159.full.pdf+html

Chaban, M. (2014, May 26) Turbines Popping Up on New York Roofs, Along with Questions of Efficiency. *New York Times*. www.nytimes.com/2014/05/27/nyregion/turbines-pop-up-on-new-york-roofs-along-with-questions-of-efficiency.html?_r=0

Lovins, A. B. (1977) *Soft Energy Paths: Towards a Durable Peace*. New York: Harper Colophon Books.

Moniz, E. (2013) www.greentechmedia.com/articles/read/DOE-Head-Ernest-Moniz-Delivers-First-Major-Policy-Address.

Rifkin, J. (2011) *The Third Industrial Revolution: How Lateral Power Is Transforming Energy, the Economy, and the World*. New York: Palgrave Macmillan.

Stephens, J. C., E. J. Wilson and T. R. Peterson. (2008) Socio-Political Evaluation of Energy Deployment (SPEED): An integrated research framework analyzing energy technology deployment. *Technological Forecasting and Social Change* 75: 1224–1246.

Stephens, J. C., E. J. Wilson, T. R. Peterson. (2014) Socio-Political Evaluation of Energy Deployment (SPEED): A Framework Applied to Smart Grid. *UCLA Law Review* 61(6): 1930–1961.

Index for Smart Grid (R)Evolution